Because the Future Matters: *Let's Stop Letting Modern Economics and Our Energy Addiction Ruin Almost Everything!*

By John M. Braden

Published by John M. Braden

Copyright John M. Braden 2015

Dedication

To the memory of my mother Joan Page (1916-2004) who, as long as she lived, showered her six children with unconditional love. She was also a peace activist.

Acknowledgements

A good number of friends and extended-family members have kindly provided encouragement, suggestions and help during the roughly forty years that I have been working—off and on—on this book. I will mention several of those people here, omitting a few who might prefer not to be linked publicly with the ideas that I am putting forward.

I will start with my brother Norman Braden. Over the years, he and I have discussed on countless occasions many of the topics covered in this book, including Hundred Percent Money in particular. "Give lots of examples!" he often said. And time and again, he referred me to relevant books, articles, Internet pieces and so on.

From Ann Braden, my daughter, I received years and years of computerized word-processing help, until finally I acquired enough basic word-processing skills to manage on my own. She also tracked down hard-to-find books for me, identified obscure quotations and helped in many other ways.

Back in the 1980s, the late Michael Denny read and commented on some early chapters of a previous version of the manuscript. I particularly appreciated his help, knowing that he was not always in agreement with my views.

Years later, Robert Snefjella went over a revised version of my manuscript with a fine-toothed comb, using coloured highlighters to categorize his comments on paragraph after paragraph. He encouraged me to re-name my basic proposal, and eventually I came up with "Intelligent National Frugality".

Towards the end of her long illness, the late Lisa Schneider also read my manuscript. I took to heart her admonition: "You use the phrase 'of course' far too often!"

More recently, Alan McNairn read and commented on a version of the manuscript much closer to the current one. Again, I was pleased to be able to mull over his suggestions, a number of which I adopted.

Other people who helped me in one way or another include George Best, Claire Baden, my brother David Braden (who is very knowledgeable about small-scale agriculture, energy-efficient housing and municipal politics), Tonny Braden, Frank Buck. my sister Gwyneth Buck, Naomi Buck, Robert Buckner, the late Gary Cooper (of Edmonton), the late Eileen Kennedy, John McColl (my cousin), the late Robert McNairn and Clare Wasteneys.

In addition, the librarians in the Tweed and the Madoc Libraries were a major help in finding reading materials which often had to be sourced through the inter-library loan service.

For the cover of my book, I am indebted to the graphic talents of Jim Smith.

I also need to acknowledge the invaluable help that I have received from Miriam Kearney and Tammy Hurrell in preparing the manuscript for the Internet and for publication as an e-book and as a print-on-demand paperback.

Lastly, I wish to thank my wife, Judith Wolfe, for helping to ensure that I had enough time to do all the reading, writing and rewriting involved. Her editing suggestions were also invariably thoughtful and helpful.

About this Book

Because the Future Matters: *Let's Stop Letting Modern Economics and Our Energy Addiction Ruin Almost Everything!* is the product of my interest in the fundamental role played by energy in all economic life.

Using straightforward language, I jump right in (notably in the first three chapters) with a set of proposals intended to go a long way towards remedying three major interrelated problems that are currently confronting Canada and the rest of the modern industrialized world: (1) increasingly serious environmental degradation, (2) the increasing likelihood of a sudden shortage of one or more key natural resources and (3) the increasing likelihood of a disastrous social breakdown.

I call this set of proposals "Intelligent National Frugality", which I often abbreviate to INF as, for example, in the phrase "INF economics".

Taken as a whole, my book is, I think, fairly original, despite the fact that, as I acknowledge, the book contains hardly any idea that has not already been put forward by someone else. I back up that assertion with a considerable number of quotations, most of which are thought-provoking in their own right.

The word "frugality" merits comment. In the opinion of a good many people, the words "frugal" and "frugality" turn today's readers and listeners right off. On the other hand, if the word "frugality" best describes the goal being pursued, then why not call a spade a spade? Also, there is primary frugality and there is secondary frugality. In my view, the kind of frugality that focuses on private individuals should be categorized as secondary frugality. Primary frugality, the heart of the matter in modern industrialized countries such as Canada, has a different focus.

Specifically, primary frugality focuses on the business world and on government. Once we succeed in establishing primary frugality, secondary frugality will no doubt follow as a matter of course. But without the former, the latter will not amount to much. Taken together, as described in this book, the two constitute Intelligent National Frugality.

What would Intelligent National Frugality look like? How would it work? How would it improve a great many features of modern industrial society? What are its ethical attractions? How compatible is it with free enterprise and/or with socialism? Realistically, could a country with as small a population as Canada's adopt INF economics before others do so? These are the principal questions that my book tries to answer.

October 4th, 2015
John M. Braden

Table of Contents

Introduction

Introduction:
Modern Energy Extravagance

The tip of the iceberg

Some news stories do more than just tell us the news. They show us the tip of the iceberg. Consider, for example, a front-page news story that appeared in *The Wall Street Journal* on August 30, 2004. According to the report, a company in Taiwan that manufactures flat screens for desktop monitors and television sets had just received delivery of some state-of-the-art manufacturing machinery from a supplier in Germany. Altogether, the machinery weighed 210 tons. How was it transported over the considerable distance from Germany to the Far East? Not by ship. Nor via the Trans-Siberian Railway. Most of the machinery got loaded onto the world's largest aircraft (an Antonov 225) and the rest onto a somewhat smaller plane (a Boeing 747). The total shipping cost amounted to half a million U.S. dollars, about twice what sea freight would have cost.

Why would the Taiwanese firm have indulged in such extravagance? It did so for an easily understandable commercial reason: The sooner the machinery in question got installed in the firm's brand new factory in Taiwan, the sooner the company could begin making huge sales of flat screens to global markets. By choosing air freight over sea freight, the company saved itself forty-five valuable days. So from the point of view of all of the parties directly concerned—the German manufacturer, the Taiwanese customer and the owners of the aircraft involved—the decision to ship by air was commercially sound.

But from the point of view of the welfare of the whole human race, including the welfare of future generations, and from the point of view of the environment, the decision was totally inappropriate. Clearly, in the overall scheme of things, it makes no sense to consume enormous quantities of non-renewable aviation fuel for the sole purpose of saving a few weeks' time in the delivery of some factory hardware.

Of what iceberg does this news story show us the tip? As I see it, we are looking at the iceberg of modern energy extravagance. I would even go so far as to call it the iceberg of modern energy insanity. For every readily visible example of energy wastefulness such as the one in this news story, there are thousands of others that we never actually hear about or read about. Nevertheless, we are all familiar with the general patterns of the modern economic world. The iceberg in question is an absolutely enormous one, and we need to stop pretending that we are unaware of its huge size.

"Insane" and "insanity" are strong words, I know. But I use them advisedly because one kind of insanity is characterized by an inability to understand, acknowledge and accept reality. Reality tells us that all of the fuel consumed in the shipping of the factory machinery by air freight to Taiwan was derived from a non-renewable and hence irreplaceable resource. On one hand, we keep telling ourselves how great our need is for energy resources. On the other hand, we keep frittering away our most irreplaceable energy resources on activities that border on the ridiculous. To me, that does not sound like sanity.

Using different but equally unflattering language, Tim Flannery makes a similar point: "But while we sit in our air-conditioned homes and eat, drink and make merry like cattle in a feedlot without the slightest thought about the consequences of our consumption of water, food and energy, we only hasten the destruction—in the long term—of our kind."

Even the titles of certain relevant books suggest the kind of insanity that I am talking about: *Our Angry Earth* by Isaac Asimov and Frederik Pohl, *Making Peace with the Planet* by Barry Commoner, *Silent Spring* by Rachel Carson and *Economic Insanity* by Roger Terry. And Andrew Nikiforuk, at the back of *Tar Sands*, sets out his "Twelve Steps to Energy Sanity".

In the case of E. F. Schumacher's classic *Small is Beautiful: Economics as if People Mattered*, its ironic subtitle speaks volumes on the point in question.

Perhaps I should express myself more gently by saying that conventional economics virtually compels sane people to engage in insane economic behaviour. In any case, most of this book focuses on what I consider to be the key questions: Where do we go from here? Why?

In my view, solutions to the problem of modern wastefulness and modern economic insanity do exist. They are not even overly complicated, at least not in their broad outlines. They do not require further technological development, further research, further economic growth, or further anything else. They require only one thing: determination.

At present, we lack that determination. Our future therefore looks quite bleak. But just as a self-indulgent individual can decide to pull up his or her socks and start behaving in an adult and responsible manner, so too can whole societies look at themselves in the mirror and say, "What we have been doing up until now is foolish beyond words! We need to make some big changes in the way we live our lives, and we need to begin doing so right now." In short, we need to make a great "Leap" onto a brand new path, if I may use Chris Turner's term.

The needed changes will require hard work. But most of us are actually quite good at hard work and are already working hard.

Hard work is not what we are afraid of. Rather, we are afraid of appearing to be stupid or ignorant or out of touch or out of date. We are afraid of being laughed at and left behind. We are afraid to say, even to ourselves, "I will not allow the current economic values of the society in which I live to dictate to me what is right and what is wrong, what is wise and what is foolish. I will listen, I will observe, and I will reflect, but if my inner self tells me that X is good and Y is bad, then I will stick with that view no matter how frequently the powers that be preach the contrary."

I can think of five different explanations, none of them mutually exclusive, for our current love affair with modern economic insanity.

Hans Christian Andersen's explanation

Back in the nineteenth century, the Danish writer Hans Christian Andersen wrote a fairy tale titled (in its English translation) "*The Emperor's New Clothes*". Long ago, the story goes, a vain and not overly bright emperor, inordinately fond of fine clothes, allowed two rogues to convince him that they could supply him with cloths and clothing not only of unsurpassed beauty but also possessing "the wonderful property of remaining invisible to everyone who was either stupid or unfit for the office he held."

Needless to say, the invisibility of the clothing stemmed from the clothing's non-existence! But because they believed the claims of the two rogues, neither the emperor himself nor any of his courtiers, nor even the general public, could bring themselves to acknowledge their apparent blindness. It was left to a small child, too young to understand adult timidity and adult social pressures, to blurt out, "But the Emperor has nothing on at all!"

To see the relevance of that fairy tale to today's world, it is only necessary to substitute "modern economics" for the vain and unclothed emperor. We are constantly being told what magnificent robes modern economics is wearing. And yet we can all point to numerous examples of modern economic foolishness and short-sightedness, including such transportation absurdities as the one described at the beginning of this chapter. That is why Tony Judt wrote, "The emperors of economic policy in Britain and the US, not to mention their acolytes and admirers everywhere from Tallinn to Tbilisi, are naked."

The "killing the goose" explanation

Long ago, some imaginative and perceptive thinker (perhaps Aesop, perhaps someone even earlier) likened one aspect of the human personality to a strange and perverse desire to kill the mythical goose that lays golden eggs. This perversity can affect individuals and it can also affect whole societies. In the latter case, we as a society get so carried away with greed and envy and thoughtlessness that we go ahead and destroy things that should have been precious to all of us: an old-growth forest together with its ecosystem, a unique human culture with its own unique language, a beautiful lake, a belt of excellent farmland, a productive fishery, a thriving human community, a heritage building, perhaps even a beneficent climate.

Thomas Berry writes as follows: "As with the goose that laid the golden egg, so the Earth is assaulted in a vain effort to possess not simply the magnificent fruits of the Earth but the power itself whereby these splendors have emerged."

The addiction explanation

Still another approach is to view our economic insanity as a kind of addiction. Just as an individual can become addicted to excessive consumption of alcohol or tobacco, so a whole society can become collectively addicted to some form of ultimately self-destructive behavior. In the case of the early inhabitants of Easter Island in the Pacific Ocean, for example, it appears (as Jared Diamond, Chris Turner, Ronald Wright and others have asserted although not everyone agrees) that the islanders' collective addiction to the continual erection of huge religious monuments led eventually to massive environmental degradation and to the community's downfall.

In the case of the modern industrialized world, our whole society has become addicted to excessive energy consumption, has it not? I single out energy consumption (the meaning of which I clarify in Chapter 9) in particular because all, or nearly all, the other resources that we consume to excess are made available to us by means of our excessive energy consumption. By putting a stop to the excesses of the latter, we will go a long way towards solving most of our other resource-consumption problems.

Interestingly, in the last ten years or so it has become socially acceptable to talk about our addiction to oil and to fossil fuels in general. But we still tend to shy away from acknowledging that we might be addicted to energy itself, i.e. to excessive consumption of energy in general. In one part of our mind, we still want to believe that virtually unlimited quantities of some new kind of energy will become painlessly available to us, provided we look in the right places and do the right research. We therefore tend to frown in disapproval when we see, for example, a book title such as *Feeding the Fire: The Lost History and Uncertain Future of Mankind's Energy Addiction* by Mark E. Eberhart.

Barbara Tuchman's explanation

In her 1984 book *The March of Folly: From Troy to Vietnam*, Barbara Tuchman focuses on government folly and government wooden-headedness, rather than on population-wide addiction. Her opening paragraph sets the stage: "A phenomenon noticeable throughout history regardless of place or period is the pursuit by governments of policies contrary to their own interests ... Why do holders of high office so often act contrary to the way reason points and enlightened self-interest suggests? Why does intelligent process seem so often not to function?"

Of particular relevance is a point that Tuchman makes about power, even though she is referring primarily to political and administrative power: "We all know ... that power corrupts. We are less aware that it breeds folly; that the power to command frequently causes failure to think ..." Tuchman does not say so, but in my view there is a close parallel between (a) the way in which having and exercising too much political power can negatively affect the judgment of political leaders and (b) the way in which having and exercising too much physical power (via gasoline, electricity and so on) can negatively affect the judgment of all of us. I will have more to say on this point in Chapter 22.

I would also like to quote the following rhetorical question presented by Tuchman as one of a great many examples of the kind of folly she is examining: "Why does American business insist on 'growth' when [such growth] is demonstrably using up the three basics of life on our planet—land, water and unpolluted air?" (Business, she notes, is not strictly government in the political sense but nevertheless represents a "governing situation".) Why indeed? And why do the rest of us go along with that insistence?

Tuchman focuses her attention on the folly of leaders rather than on that of the man and woman in the street. I myself would argue, however, that the responsibility for modern economic insanity is essentially split fifty-fifty between leaders and followers. In other words, I do not disagree at all with Tuchman's thesis about government folly, but surely in a democracy we the people have considerable choice as to who will lead us. If we had truly wanted to turn our backs on modern economic insanity, we could fairly easily have so informed our politicians long ago and then voted accordingly.

(In that regard, we could in my view improve the democratic nature of our political elections in a very simple way: We could institute a general rule whereby on every election ballot, directly beneath the names of all the election candidates, would appear the words "NONE OF THE ABOVE" together with a corresponding space for the voter's "X". All votes for "NONE OF THE ABOVE" would have to be officially counted and the total officially published. Imagine the democratic ferment that would occur if more than fifty percent of the voters were to select "NONE OF THE ABOVE"! I would consider this improvement to be absolutely essential if we ever decided to make voting compulsory.)

The "tacit agreement" explanation

Years ago, we, the public, entered into a tacit agreement with the economics profession. "You tell us how the economy ought to work, including what its specific goals should be," we said to the profession, "and we'll leave the whole field to you and your expertise. Just make sure that we all get richer and richer!" Understandably but sadly, most of the economics profession accepted this offer and gave us what they thought we wanted: a recipe that would supposedly enable our economy to keep on growing forever. The rest, as the saying goes, is history.

Our need for change

Regardless of how each of us chooses to analyze the problem of modern economic insanity, we need to change our collective economic behaviour. Otherwise, as countless writers and others have been warning, we run the risk that one or more of three interrelated disasters might occur (either gradually or suddenly):

- Disastrous environmental degradation.
- A disastrous shortage of one or more key natural resources.
- A disastrous social breakdown.

In this book, I will not be spending much time discussing the details of the above three disaster risks. Many other books have already done so. I am much more interested in the question of how those risks can be intelligently minimized. Put another way, I hope that through the combination of wisdom and good luck, no one ever finds out which of the above three disaster risks was fated to actually materialize ahead of the other two.

No doubt there are several different ways of putting a sensible end to modern economic insanity. The proposals that I put forward in this book are the best ones that I have been able to come up with. But I am not claiming originality for any of them. Over the years, I have done enough reading to know that there is hardly a single idea, if any, in this whole book that has not already been voiced elsewhere by someone else, either recently or many years ago or both. Various quotations that appear herein serve to back up that assertion. (In my view, they are also interesting and thought-provoking in their own right.)

As an overall name for the set of proposals that I am putting forward, I have chosen the term "Intelligent National Frugality", which I will often abbreviate to "INF". The word "frugality" stands in stark contrast to the ideology of perpetual economic growth that we have all been brought up to believe in so unrealistically and that I will be discussing in Chapter 13. The word "national" indicates that the frugality we need should extend throughout the whole of our society, including all businesses, all professions, all organizations and all levels of government, rather than being confined to how you and I live our own private lives. And the word "intelligent" is intended to suggest that some conceptions of frugality probably make more sense than others.

By itself, Intelligent National Frugality will not solve the totality of our economic and social and environmental problems. No single set of economic proposals will ever do that. There will always be a need for adaptations, regulations, prohibitions, restrictions, social experiments and so on. In addition, quite apart from matters of government policy, there will always be a need for strong community spirit and goodwill. Once implemented, however, INF economics will in my view do two things. It will methodically remove much of the insanity, absurdity and environmental abuse that permeate modern economic life. And it will provide very real encouragement to all those wishing (a) to raise the intrinsic value that we place on the whole family of human beings (past, present and future) and on Gaia and (b) to lower the value that we place on the almighty dollar. As far as I know, no one has ever made a similar claim on behalf of conventional economics.

The Gaia theory, incidentally, is most closely associated with the names of James Lovelock and Lynn Margulis. Lovelock's own summary description of the theory is as follows: "A view of the Earth that sees it as a self-regulating system made up from the totality of organisms, the surface rocks, the ocean and the atmosphere tightly coupled as an evolving system. The theory sees this system as having a goal—the regulation of surface conditions so as always to be as favourable as possible for contemporary life."

Part 1: The Basic Proposal

1

Human Energy

"There are both upper and lower limits to the rate of [human] energy expenditure. By running upstairs three at a time, or chopping wood furiously or playing tennis with all one's might, energy can be expended at the rate of 10 kilocalories per minute over a short period of time ... On the other hand, by suspending all voluntary activity, the rate of energy expenditure can be cut down to a fifteenth that quantity, but no lower."

Isaac Asimov (1962)

"Of all the organs in the body, the brain is the greediest in its fuel consumption. It burns oxygen and glucose at ten times the rate of all other body tissues at rest. In fact, the brain uses up so much energy that it dies if deprived of oxygen for only a few minutes."

Susan A. Greenfield (1997)

A request

Please read this chapter and the next one carefully. Together, they form the foundation that supports most of the rest of the book.

Human energy: definition and description

In recent years, we have all become fully aware of the crucial role played by energy in the modern economic world. Energy of one kind or another has in fact always been crucial in human economics. As far back as 1926, Frederick Soddy wrote the following words: "If we have available energy, we may maintain life and produce every material requisite necessary. That is why the flow of energy should be the *primary* concern of economics." (Emphasis added.)

There is one particular kind of energy that we tend to overlook when thinking about economics: human energy. If this book were a novel, however, "human energy" would be its hero.

What do I mean by human energy? A simple answer is that human energy refers to the energy that flows through human beings. But for those who (like myself) prefer to be given precise definitions and descriptions, I offer the following:

First of all, I have chosen to define human energy in such a way that comparisons with other forms of energy, notably nuclear and fossil-fuel energy, are facilitated. Accordingly, I use "human energy" as a shorthand way of saying "one particular portion of the energy output of human beings, namely the portion that gets allocated to economic activities". Now for some additional comments and clarifications:

1. I am talking here about energy output, not energy input.
2. I am talking here about only a portion of the energy output of a human being, not the total amount.
3. The portion in question consists of the portion allocated to economic activities. All other portions are excluded.
4. For human beings, just as for gasoline engines and so on, total energy output can never exceed total energy input. Both everyday experience and the laws of science confirm that fact. Moreover, losses are inevitable. Consequently, the amount of useful energy output will always be (a) significantly less than the total energy output and hence (b) significantly less than the total energy input.

5. Energy input therefore deserves attention. For human beings, all our energy input comes essentially from the chemical energy contained in the food we eat. In the case of the average, properly fed, human adult, the amount of food energy taken into the human body every day averages approximately 2,000 kilocalories. (In normal conversation, we refer to food kilocalories as "calories". But in physics and chemistry, a "calorie" is actually 1,000 times smaller than a food kilocalorie.)

6. Energy losses are unavoidable in human beings (and other living organisms) just as they are in mechanical engines. Consequently, the useful portion—including the portion allocated to economic activities—of the energy output of the average person will always average less than approximately 2,000 kilocalories per day.

7. The exact portion (of any particular person's total energy output) that gets allocated to economic activities is difficult, if not impossible, to measure precisely. But that does not matter and is not important. What does matter is the fact that the portion in question will always be less than 100% and hence in the average case will always consist of less than approximately 2,000 kilocalories per day. In other words, we know the approximate upper limit (2,000 kilocalories per day), even if we do not know the exact amount.

8. QUESTION: You talk about the "allocated to economic activities" portion of the energy output of a human being. Can you give us some examples of the kinds of activities you have in mind? ANSWER: Certainly, madam. Examples would include the kinds of activities (a) that employers pay their employees to engage in, (b) that a corporate board of directors would expect its chief executive officer to engage in, (c) that clients pay their

lawyers and accountants to engage in, (d) that patients (or their health insurance companies or their governments) pay their doctors and dentists to engage in, (e) that entrepreneurs and the self-employed engage in for the purpose of earning a living and (f) that self-sufficient persons engage in for the purpose of producing what they and their families consume.

9. Note that some of these activities are mostly physical whereas others are mostly mental. They all, however, fall into the category of the "allocated to economic activities" portion of the energy output of human beings. So they all fall into the category of "human energy" as that term is used in this book.

10. The physical energy output of a person playing hockey strictly for fun does not meet my economic definition of human energy, whereas the physical energy output of the very same person playing hockey as a professional athlete does qualify. Similarly, the mental energy output of a college student writing an algebra examination does not meet the definition, whereas years later when the student has become a professional engineer doing mathematical work on the design for a new railway bridge, his or her mental energy output in connection with that bridge project does qualify.

11. The most obvious way in which we allocate human energy to economic tasks is via our muscles. We lift, we carry, we run, we walk, we push, we pull, we dig, we throw, we knead, we haul. As I have already mentioned, however, muscular effort makes up only half of the story. The other half consists of mental activity. We think, we plan, we design, we read, we imagine, we calculate, we supervise, we observe, we communicate, we teach, we investigate, we judge.

12. In most cases, human beings engaged in doing work are actually expending a blended combination of physical energy and mental energy. The proportions of the two will of course depend upon the kind of work involved. The physical component in the case of lumberjacks and ballet dancers will be much greater than in the case of poets, librarians and newspaper editors. When we hire other people to work for us, we are in a sense purchasing a certain portion of their physical energy output and a certain portion of their mental energy output. Similarly, when we work on our own behalf, we allocate to the task at hand a certain portion of our physical energy output and a certain portion of our mental energy output.

Human dignity and the human spirit

Although human effort can be analyzed in energy terms, no economic system could be described as civilized if it treated human energy as just another form of energy. That is why our society rejects, or at least is supposed to reject, slavery, forced labour, child labour, underpaid labour and unsafe and unhealthy working conditions. Any worthy system of economics must respect human dignity and the human spirit. In my view, conventional economics does not pass that test. I agree completely with Wendell Berry's observation: "Our time [i.e. our era] is characterized as much by the abuse and waste of human energy as it is by the abuse and waste of fossil fuel energy." The same point is made by Paul Hawken, Amory Lovins and L. Hunter Lovins in their book *Natural Capitalism*: "A society that wastes its resources wastes its people and vice versa."

Two remarkable "attachments"

Unlike electricity and the various fossil fuels, human energy comes complete with two remarkable "attachments": the human brain and the human body. It comes complete, in other words, with what is perhaps the world's most sophisticated computer as well as with what is perhaps the world's most versatile piece of machinery. Most other forms of energy are quite naked. They therefore tend to be of little value in the absence of (a) some sort of hardware such as an electric motor or an internal-combustion engine or a soldering iron and (b) some sort of decision-making entity such as a computer, a thermostat or a human supervisor or operator.

The technological and environmental implications of the two "attachments" possessed by human energy are of major importance. In most situations, far less machinery and equipment and far less environmental stress are involved in the effective use of human energy than in the effective use of nuclear or fossil-fuel energy.

An axe for splitting firewood, for example, is a simple device involving only a small amount of technology in its manufacture and involving virtually no environmental stress in its use. But a hydraulic wood-splitter powered by a gasoline-fueled engine involves both a significant amount of machinery and a significant amount of air pollution and carbon dioxide (CO_2) production. Both the axe and the hydraulic wood-splitter do the same job. The attraction of the hydraulic wood-splitter lies in the fact that its use requires less human effort and less manual skill than does the axe.

Energy wastage

Imagine a truck whose engine must for some reason be kept running at all times and whose fuel tank must therefore get continually refilled. Under such circumstances, it would clearly be foolish and wasteful not to find some useful work for the truck to perform. As that analogy suggests, the modern world's current pattern of widespread unemployment, underemployment and inappropriate employment can be criticized not just as a moral shame but also as a foolish waste of high-quality energy.

Back in 1933, George Orwell made that very point. Discussing the effect of a law that in his opinion virtually forced tramps in England to walk endlessly from town to town, Orwell wrote: "There must be ... several tens of thousands of tramps in England. Each day they expend innumerable foot-pounds of energy—enough to plough thousands of acres, build miles of road, put up dozens of houses—in mere, useless walking."

One key fact about human energy is that in a civilized society someone has to pay for it, either in cash or in kind, regardless of whether it gets put to productive use or not. We all need food, clothing and shelter. Somehow, in one way or another, nearly all of us do get the food, clothing and shelter we need. If I am not able to pay for mine, then either society as a whole or my family or some charity will make the payment on my behalf. Someone will pay the upkeep cost of my human energy no matter what. (Sadly, there are exceptions in today's Canada, notably in cases where unsheltered persons end up freezing to death during the winter.)

I would add that both the moral shame and the waste of energy can extend to cases where physical or mental handicaps are involved. Nearly everyone has the ability to allocate a significant portion of his or her human energy output to some

21

useful purpose, especially if help from a competent human supervisor is both available and affordable. No economic system can be considered either worthy or energy-efficient if it rejects reasonable offers of human energy from the handicapped.

Energy competition

Being a form of energy, human energy often finds itself in direct economic competition with other forms of energy.

Sometimes, this competition is quite obvious. A human welder on an assembly line, for example, may have to compete with a robot. A night-time security guard may have to compete with a closed-circuit video system. A bank teller may have to compete with an automated teller machine. An architectural draftsperson may have to compete with the energized software of a computer. In each case, human energy is competing directly with the energy that powers the robot, the video system, the ATM, the computer and so on. More precisely, the human energy is competing with the combination of (a) the energy used to manufacture the robot, (b) the energy used to power the robot and (c) the energy used to perform repairs and maintenance on the robot.

Sometimes, however, the competition is less obvious. Consider, for example, the case of a senior business executive who flies frequently back and forth between his or her head office and the firm's five main factories, all quite distant from each other. Here the energy competition is easily overlooked. But suppose air travel were unsafe or technologically impossible or prohibitively expensive. In that case, each of the five factories would no doubt be assigned its own senior executive officer. Or each factory might end up having a completely separate owner. The energy competition here is thus between (a) one unit of human energy combined with a very large quantity of aviation-fuel energy and (b) five units of human energy combined with little or no aviation-fuel energy.

Typically, as in the example just given, the competition is between different combinations of energy types. And each energy combination participating in the competition will usually contain at least some minimum quantity of human energy. In many situations, for example, there is little point in having a closed-circuit video security system unless some human energy is available to keep an eye on what the video screens are actually showing.

At one extreme, the energy combination might consist exclusively of human energy. The services of a babysitter who walks over to one's home from next door, for example, might well involve no extra heat, electricity or fuel at all. At the other extreme, the quantity of human energy required to fly a jumbo cargo jet aircraft across an ocean represents only a tiny fraction of the total energy consumed by such a flight. And lying in between those two extremes are activities where substantial quantities of both human and inanimate energy are involved, such as the construction of a high-rise office building.

What factor determines the winner in competitions between various energy combinations? Price does, in most cases. The energy combination having the lowest overall cost will normally get selected in preference to all other possible energy combinations. Hence if competent human energy is priced at an attractively low level— attractively low from the point of view of the purchaser—then the energy combination chosen is likely to consist largely of human energy. If on the other hand fossil-fuel energy is priced at an attractively low level, then the energy combination chosen is likely to consist predominantly of fossil-fuel energy. In short, energy choices depend for the most part upon the relative prices of the various kinds of energy.

The price of human energy

For most kinds of energy, the price—expressed in dollars per unit of energy—can be calculated quite easily. If one knows the price of a litre or gallon of gasoline and the number of energy units (joules or British thermal units) contained therein, simple arithmetic will yield the price of gasoline energy. In the case of electrical energy, no calculation of any kind is required since electricity is already priced in kilowatt-hours and the kilowatt-hour is a unit of energy.

Any energy unit can be converted by simple arithmetic into any other energy unit, e.g. kilowatt-hours to British thermal units, or joules to calories, or foot-pounds to joules. One kilowatt-hour, for example, equals approximately 800 kilocalories. Hence one can readily compare the prices of most kinds of energy.

The price of human energy, however, cannot easily be expressed precisely in dollars per unit of energy. The problem is that there is no easy way to calculate what portion of one's daily intake of food energy ought to be allocated to the work for which one is being remunerated.

If I earn $100 for a full day's work and consume 2,000 kilocalories of food energy per day, then my energy output will cost my employer 5 cents per kilocalorie at the very least ($100 divided by 2,000 kilocalories). That would be the cost in the extreme case where all 2,000 kilocalories were somehow able to be allocated exclusively to my paid work. But if only, say, 200 kilocalories were able to be so allocated, then my energy would be costing my employer 50 cents per kilocalorie ($100 divided by 200 kilocalories). One could perhaps conclude that my energy is costing my employer more than 5 cents per kilocalorie but less than 50 cents per kilocalorie. Still, that information is not very precise.

It turns out that the question of how to calculate the precise cost of human energy is an academic one and hence need not concern us further. In the real world, a potential employer of human energy is only interested in the comparative cost of various options. Will the use of a robot, for example, cost more money or less money than the use of the human energy for which the robot might be substituted? An accountant can easily make a direct comparison between those two costs without having to analyze human energy in terms of cents per kilocalorie or dollars per kilocalorie.

Numerically speaking, then, the price of human energy is not important. Conceptually speaking, however, its price is extremely important. More precisely, the relationship between the price of human energy and the price of non-renewable forms of energy is extremely important. The whole theme of this book centres around my contention that, in the wealthy countries of the world at the present time, the relationship between those two prices is terribly unsound. Human energy, I argue, is at present vastly overpriced, while fossil-fuel energy and nuclear energy are vastly underpriced. As a result, our existing economy is virtually guaranteed to perform badly, notably with regard to human employment, resource conservation and the environment. Put into stronger language, our existing economy is insanely unsound.

When I say that human energy is at present vastly overpriced, I am generalizing. Clearly, some human energy costs no money at all. Examples include cases where one builds one's own house, repairs one's own car, grows one's own food, prepares one's own income tax return, and so on. Other examples include cases where one does volunteer work at one's local hospital, food bank or fund-raising supper. But in the normal commercial situation where a person receives a wage or salary or commission for work performed, the cost of the human energy involved is extremely high under conventional economics in the advanced industrialized countries of the world. That key fact underlies just about everything I say in this book.

If the cost of human energy is already extremely high in the case of people earning a normal income, what about those individuals whose remuneration exceeds a million dollars a year? The price paid for their human energy can only be described as astronomical. Just imagine what the cost per kilocalorie must be in the case of a corporate chief executive officer earning $10,000 per day. Even if all 2,000 kilocalories of the CEO's daily energy intake could somehow be allocated to his or her economic output, the cost of the human energy involved would still amount to $5 per kilocalorie. As mentioned, one kilowatt-hour is approximately equal to 800 kilocalories. So if one kilocalorie costs $5, then 800 kilocalories cost $4,000, which means that one kilowatt-hour's worth of the CEO's human energy costs $4,000. Compare that figure with the cost per kilowatt-hour on your present electricity bill!

Actually, in wealthy countries such as Canada, even minimum-wage human energy is far more expensive, relatively speaking, than one might think. Our blindness here is due to the fact that we have become thoroughly accustomed to consuming huge quantities of non-renewable energy without being fully aware of the numbers involved. As a society, we consume only tiny quantities of human energy compared with our consumption of non-renewable energy. We therefore tend not to fully realize that under conventional economics there is an enormous difference between the price of human energy and the price of non-renewable energy.

Chris Turner has made the same point by calculating how much a barrel of crude oil would cost if it were priced at the same number of dollars per unit of energy as we currently price minimum-wage human energy. His answer is that a barrel of oil would cost $138,000! In his book *The Leap*, Turner explains clearly how he arrives at that answer.

The right of first refusal

In the next chapter, I will be describing some key mechanisms for converting our society into something much more in keeping, I believe, with our basic values and our legitimate aspirations. Here I would just like to sum up very briefly the difference between (a) the way we ought to deal with human energy and (b) the way we in fact deal with human energy at present. The difference could not be more striking.

Ideally, we need an economic system in which human energy gets singled out by being granted a special privilege, namely the right of first refusal. When there is a task to be performed, such a system will turn first to human beings and say, "If you want this job and are capable of doing it well, then you may have it and you will receive reasonable remuneration in exchange. We will not offer the job to other forms of energy unless and until you decline it. In short, we are granting you the right of first refusal." Note that everyone gets offered a wide range of jobs, while no one gets forced to choose any particular job. We all consume, so it makes sense to let us all produce. And ideally we should all be able to produce in such a manner that we make a real contribution to society and, at the same time, we fully develop our human and creative potential.

Conventional economics takes a very different tack. "Be warned, everyone!" it says. "None of the jobs that you are currently doing are permanently reserved for human energy. In fact, fossil-fuel energy and nuclear energy are much to be preferred to human energy. Accordingly, all that stands between you and unemployment is a little technological development. At this very moment our engineers and inventors are working full-time to come up with technology capable of harnessing inanimate energy to your current jobs. So be warned! No matter how fervently you may wish to keep your present job and no matter how expertly you may be performing it, most of you are destined to be replaced. The fact that your human energy may then go to waste does not cut any ice at all. Nor does the fact that the required quantities of replacement energy may be very large. Under present circumstances, we pay little attention to overall quantities of energy consumed, except in those cases where significant amounts of money can be saved."

In my view, Intelligent National Frugality (INF) economics will succeed in correcting this terrible flaw in conventional economics. It will do so primarily by means of its price structure, which will be revolutionarily different from the one with which at present we are all familiar.

2

Intelligent National Frugality (INF) Economics

"However, the shift to lower energy use could be produced more simply and consistently by a tax on energy. This process would work through the energy market: if energy were taxed as it left the ground, well-head, or hydropower dam, or as it was imported, on the basis of actual energy content, then the effects would filter through the economy and appear as increases in the cost of consumer goods and services."

Bruce Hannon (1977)

"So why not shift the tax base from value added (earned income) and on to that to which value is added (natural resource throughput)? ... This seems such an obvious improvement that one wonders why economists remain so in thrall to value-added taxation."

Herman Daly (2011)

As I suggested in the previous chapter, conventional economics places far too high a price on human energy. More precisely, it places far too high a price on human energy as compared with the price of fossil-fuel energy and nuclear energy. Or, put the other way around, conventional economics places far too low a price on fossil-fuel and nuclear energy as compared with the price of human energy.

Once one accepts that particular diagnosis, the broad outline of the remedy becomes obvious. We need to lower the price of human energy and/or raise the price of fossil-fuel and nuclear energy. We need to bring about a harmonious balance between the two. I would define a harmonious balance here as a balance providing adequate welfare and opportunity for everyone: for future generations as well as for ourselves.

Let's turn to specifics

In any discussion of prices, the topic of taxation is always relevant. If something gets newly taxed or gets taxed at a higher rate than before, the total price that the buyer has to pay goes up. Similarly, if a tax is lowered or abolished, then the total price to be paid by the buyer goes down. (I am assuming here that the seller does not change his or her own pre-tax price.)

I begin, then, with the hypothesis that perhaps our taxation system is responsible for the lack of a harmonious balance between the price of human energy and the price of fossil-fuel and nuclear energy. Perhaps our taxation system overtaxes human energy and undertaxes fossil-fuel and nuclear energy. Put even more bluntly, perhaps our taxation system essentially taxes the wrong kind of energy. If so, then no wonder our economy seems insane.

On what kind of energy does it make sense to levy taxes? On all kinds equally? On all kinds but not equally? Or only on some kinds but not others?

A general principle of taxation policy can help us here. According to this principle, whenever taxation revenue is required, one should look first at "bads", i.e. at activities that society wishes to discourage. If at all possible, one should refrain from imposing taxes on "goods", i.e. on activities that society wishes to encourage.

Applying that general principle to energy taxation, one can easily understand the argument that human energy in the modern world ought not to be taxed at all. We want to encourage, not discourage, the reasonable utilization of human energy. We want to encourage employment, self-employment and work opportunities in general. Why place obstacles in the path of such obvious "goods"?

One can also easily understand the arguments in favour of the taxation of fossil-fuel energy resources and nuclear energy resources. Those resources are not just finite in quantity but also non-renewable. Future generations would no doubt be grateful if we left a reasonable amount of oil and coal and so on behind for them. The environment would also be grateful, since the large-scale extraction and utilization of such resources is seldom environmentally benign. And we ourselves might appreciate living in a world characterized by much more economic sanity than at present.

Based on all of this reasoning, Intelligent National Frugality (INF) economics begins, conceptually speaking, by wiping the taxation slate clean. With the exception of local property taxes, it cancels all existing taxes. In particular, it cancels the personal income tax, the corporate income tax, all general sales taxes and all taxes on "added value", such as Canada's Goods and Services Tax.

The reason for cancelling all these taxes is not that they tax human energy exclusively. None of them in fact do that. But they all do tax some human energy some of the time. INF economics takes the position that, as a general rule, human energy should never be taxed at all. Hence cancellation is in order.

As with existing taxes, so with existing subsidies. At the present time, most modern industrialized countries are providing several subsidies, some direct and some indirect, to fossil-fuel energy and nuclear energy. All such subsidies need to be terminated. The idea is to start off with a totally clean slate as regards both energy taxation and energy subsidization.

The next step is to replace all the lost taxation revenue. INF economics accomplishes this by instituting a special kind of energy tax. This particular energy tax is incapable of ever taxing human energy. Instead, it gets levied exclusively on the two main kinds of non-renewable energy, namely fossil-fuel energy and nuclear energy.

Nuclear energy will be discussed in Chapter 8. The only point I wish to make here is that the proposed new tax would almost certainly render nuclear power prohibitively expensive. In practice, therefore, a society operating under INF economics would have made the decision to stop using commercial nuclear power altogether. That being the case, the tax on nuclear energy would not generate any significant revenue at all.

With nuclear energy essentially out of the picture, fossil-fuel energy becomes the main focus of the new tax. Fossil-fuel energy consists of the chemical energy found in three broad categories of energy resources: coal, oil (petroleum) and natural gas. For our purposes here, those three categories include all the different kinds and grades of coal, all the different kinds and grades of crude petroleum oil (including the bitumen contained in what are known as oil sands or tar sands), all the different fuel chemicals found in deposits of natural gas (including not just the methane but also the so-called "natural gas liquids" such as ethane, propane, butane and so on), and all the natural gas obtained from unconventional sources (such as coal-bed methane) as well as from conventional sources.

Just exactly what kind of a tax does INF economics place on fossil-fuel energy?

ANSWER: I would call the tax a true energy tax because it would be expressed in dollars per unit of energy. It would thus be totally independent of the pre-tax dollar value of the resource being taxed.

The tax on a barrel of crude oil, for example, would be calculated by multiplying the rate of the tax (expressed in dollars per unit of energy) by the total number of units of chemical energy contained in that barrel of crude oil. No account would be taken of such matters as whether the oil had been easy or difficult to bring up to the surface. Nothing but the actual energy content of the oil in question would be relevant.

The rate of the tax would be the same for all the different kinds and grades of fossil-fuel energy resources. Thus the amount of tax payable on a tonne of high-energy anthracite coal, for example, would be higher than that payable on a tonne of low-energy lignite coal. But the rate itself, expressed in dollars per unit of energy, would be the same in both cases.

Upon whom would liability to pay the tax fall?

ANSWER: The person or corporation responsible for bringing the resource into the national economy, either by bringing it up out of the ground or else by importing it, would immediately become liable to pay the tax.

What about oil spills, natural gas leaks, flared-off natural gas and so on? Would such unmarketed energy resources be taxed as well?

ANSWER: Yes, they would. The rule here would be the same as in a china shop where the sign says: "If you break it, you buy it!" Translation: "If you release the resource from its underground location, or if you allow it to escape, you pay the tax on it!"

Electricity does not seem to be covered by this new tax. Why is that?

ANSWER: Electricity does not constitute a source of energy. Rather, just like the chemical energy contained in hydrogen gas (H_2), electricity is best described as a form of energy into which certain other forms of energy can be converted. Nevertheless, the cost of producing electricity would in many cases be affected substantially by the new tax. In particular, any and all inputs of fossil-fuel energy used, either directly or indirectly, in the generation of electricity would have been taxed as soon as they entered the national economy. And any electricity generated from expensive inputs would naturally end up being expensive itself.

Note that, with respect to electricity, INF economics differs somewhat from the Bruce Hannon quotation appearing at the beginning of this chapter. The INF tax would not apply to "energy as it left the hydropower dam", since falling water does not constitute a non-renewable energy resource. A case could be made, however, for imposing a partial INF tax on all future electricity generated by large, previously built, hydroelectric power stations. Such a tax would reflect (a) the fact that the original construction of the facilities in question involved substantial inputs, direct and indirect, of fossil-fuel energy and (b) the fact that those inputs predated the imposition of the INF tax and hence did not cost very much.

Incidentally, the Hannon quotation shows that the idea of taxing "actual energy content" has been in circulation at least since 1977.

Would the INF tax be passed on from the original resource extractor to the ultimate consumer of the taxed energy?

ANSWER: Yes, it would. I have already mentioned how this would happen in the case of fossil-fuel energy inputs used in the generation of electricity. In the same way, the tax would be reflected in the retail price of such fuels as gasoline, diesel fuel, natural gas, propane and so on. In fact, the retail prices of all goods and services throughout the whole economy would reflect the tax to the extent that any inputs of fossil-fuel energy were involved. Working together, in other words, the tax and the marketplace would distribute an appropriate portion of the tax burden to every relevant purchase in the whole economy.

At what rate should this new tax be set?

ANSWER: The rate should be set high enough so that the tax generates nearly all the revenue required by the various levels of government above the municipal level. In Canada's case, this means that the federal, provincial and territorial governments would obtain nearly all their revenue from the new tax. Calculation of the tax would involve three steps: (a) a decision as to how much annual gross revenue was desired from the tax, (b) an estimate of the country's total annual consumption of energy (measured in energy units) from fossil-fuel energy resources and (c) a simple division of (a) by (b). At regular intervals, the rate could be adjusted, if necessary, so as to take into account any budget deficits or surpluses from previous years. Note that the tax is revenue neutral in the sense that it is not intended to increase total government revenue but rather to change the way in which nearly all government revenue (except at the municipal level) is obtained.

What is the significance of the term "nearly all" which appears three times in the answer to the previous question?

ANSWER: The term "nearly all" is intended to leave room for special categories of government revenue, such as (a) fines levied against criminal and other offenders, (b) application fees and user fees of various kinds, (c) revenue from government investments and government-owned enterprises, (d) proceeds from asset sales and (e) revenue from any taxes (wealth taxes or inheritance taxes, for example) having as their specific purpose a reduction in the inequality of wealth in society as a whole.

Would the personal income tax be eliminated entirely?

ANSWER: Ideally, yes it would. But in practice, even under INF economics, a government would be perfectly free to conclude that the degree of wealth inequality in our society was excessive. Having reached that conclusion, the government could then take steps to deal with the problem. It could impose, for example, an income tax on all incomes considered to be excessive. Alternatively or in addition, some kind of wealth tax or inheritance tax might be considered appropriate. But for the vast majority of the population, if not for everyone, the personal income tax would no longer exist.

Although I might be mistaken, I suspect that under INF economics it would be extremely difficult for any individuals to direct into their own pockets the huge salaries and commissions and bonuses that we read about so often in today's newspapers. It is one thing to become exceedingly rich by exploiting, directly or indirectly, an abundance of cheap energy. But it is quite another thing to become exceedingly rich when most energy is either diffuse (sunshine) or revolutionarily expensive (fossil-fuel energy) or else possessed of a truly independent body and mind of its own (human energy).

Would the proposed new tax not impose unacceptable hardship on people with very low incomes?

ANSWER: Yes it almost certainly would, if steps were not taken to deal with the problem. Such steps will be discussed in Chapter 4. First, however, I would like to paint an overall picture in Chapter 3 of what our society might look like under INF economics.

Clarification concerning the source of fossil-fuel energy

There is some scientific uncertainty as to the original source of the chemical energy contained in the Earth's fossil fuels. This energy may have all been derived from solar energy, via living organisms, eons ago. Alternatively, as explained by E. C. Pielou, some of the Earth's crude oil and natural gas "may be as ancient as the solar system itself and may never have been incorporated in living organisms." Although interesting, this scientific uncertainty has no bearing on economics. It is relevant, however, to my next clarification.

Clarification concerning the term "solar-derived"

In this book, simply for convenience, I exclude all fossil-fuel energy from the meaning of the term "solar-derived energy". If I did not do so, I would have to write "solar-derived energy not including fossil-fuel energy" far too many times. The key reason for separating fossil-fuel energy from "solar-derived energy" relates to age. The vast amount of chemical energy stored in fossil-fuel energy deposits has been stored there for eons, regardless of where it came from originally. Contrast those eons with the mere centuries (or so) over which the solar-derived chemical energy in our oldest trees and our oldest wooden artifacts has been stored.

3

A Peek at the Future

"Decentralization, non-hierarchical forms of organization, recycling of wastes, simpler living styles involving less-polluting "soft" technologies, and labor-intensive rather than capital-intensive economic methods are possibilities only beginning to be explored."

Carolyn Merchant (1980)

"A sane civilization ... would have more parks and fewer shopping malls; more small farms and fewer agri-businesses; more prosperous small towns and smaller cities; more solar collectors and fewer strip mines; more bicycle trails and fewer freeways; more trains and fewer cars; more celebration and less hurry; more property owners and fewer millionaires; more readers and fewer television watchers; more shopkeepers and fewer multinational corporations; more teachers and fewer lawyers; more wilderness and fewer landfills; more wild animals and fewer pets."

David W. Orr (1994)

This chapter takes the form of an imaginary letter written to the present generation of Canadians by one of our descendants. Much has changed as a result of Canada's adoption of Intelligent National Frugality (INF) economics. Here is the letter:

Dear Ancestors,

Please accept greetings from the future!

Although our society is continuing to evolve and to change, I think it is fair to say that our transition to Intelligent National

Frugality (INF) economics has now by and large been completed. I will therefore try to describe for you some of the more visible features of our society that distinguish it from your own.

As you read these paragraphs, please keep in mind the key difference between your approach to economics and ours. That difference centres around taxation. No one here is under any absolute obligation to pay any taxes at all, apart from the obligation of landowners to pay municipal taxes. Anyone who earns any money is perfectly free to keep one hundred percent of that money if he or she so wishes. And that is true for businesses as well as for individuals.

On the other hand, anyone who purchases, either directly or indirectly, any fossil-fuel energy has to pay quite a high price per unit of fossil-fuel energy purchased. Thus, one is not taxed on the basis of how much one earns or how much one spends. Instead, one is selectively taxed, in effect, on the basis of all the fossil-fuel energy inputs, if any, included in the various items (fuels, electricity, goods, services) that one chooses to buy. For taxation purposes, we make no distinction between (a) cases where the coal or oil or natural gas is used for energy purposes and (b) cases where the resource is used as a feedstock for the manufacture of lubricants, plastics and other chemical products.

In theory, we also tax nuclear energy and we do so at the same rate as for fossil-fuel energy. But in practice, this means that we simply make no commercial use of nuclear power at all.

Because our taxation system is so different from yours, our prices are also very different from yours. In my view, you would not find it helpful if I were to state our current price for a cup of coffee or a pair of shoes or a three-bedroom house. The numbers would not mean much for two reasons. You are unfamiliar with our life-styles and values and, even more important, you are unfamiliar with all the other prices that form the background against which a single price for a single piece of merchandise can be meaningfully understood. So instead of overloading you with potentially misleading numerical information, I will try to paint a general picture for you in words.

Generalizations

Here, first of all, are some generalizations. In the opinion of most of my fellow citizens and myself, our country can be described as decidedly frugal and yet at the same time reasonably prosperous. We work hard when hard work is called for, but not every waking moment of every day. We try to give everyone an opportunity to live a satisfyingly productive life and we try to achieve economic fairness for the whole of our society, all the while maintaining a high standard of democracy. In the area of resource conservation and environmental protection, we have made considerable progress. And with each passing decade, we feel that our sense of community, both at the local level and in the broader context, is growing stronger. We still have weaknesses and failings, of course. But all in all, we are in fact quite proud of our country.

Analytically speaking, the chief economic difference between our society and yours can be summed up in the words "energy efficiency" and "frugality". We use energy far more efficiently than your society ever did and we have also become quite frugal not just in our individual lives but also in our business activities and government activities. We have been enabled and encouraged to make these changes primarily by the financial incentives and disincentives that are built into INF economics.

For the most part, the technology that we use was available to your society as well. Admittedly, we have been able to develop a certain amount of useful new technology in a few areas. But that has not been crucial. Most of our increased energy efficiency and frugality can be attributed to our technological choices rather than to our new technological inventions. In turn, those choices can mostly be attributed to our revolutionary new price structure.

Refrigeration and clothes-drying

Let me give you a simple example. The average household refrigerator today is far more efficient than was the case in your day. Without developing any new technology, we have simply made generous use of insulation and we have located the heat-producing components as far as possible from the cold interior of the appliance. We have also borrowed an old idea and made it the norm: In essence, the refrigerator in most homes today is located in a "pocket" or alcove built into one of the main walls of the kitchen. The door and contents of the refrigerator continue to be accessible from the kitchen in the normal manner. But the rear of the appliance, while sheltered from rain and snow, is exposed directly to shaded outdoor air at all times. As a result, much of the needed cooling is provided by the outside air at no cost.

This system does not provide any benefits during the hot summer months. But it definitely does during the long winter months and during early spring and late autumn. Moreover, the further north one goes and the longer the winter becomes, the greater the annual number of months of benefit.

Nor is that the whole story. Refrigerators that are cooled by outdoor air during the colder months of the year lend themselves particularly well to the use of solar-generated electricity during the hot summer months. This is because (a) a great deal more solar energy is available in summer than in winter and (b) there is much less demand for electricity for household lighting purposes in summer than in winter. Hence for a great many of today's households, significant amounts of refrigeration electricity are provided during the summer months by panels of photovoltaic cells that face the sun. In some cases, these panels belong to, and are located at, the household in question. In others, the panels are located elsewhere and supply power to the local grid.

A much simpler example of increased energy efficiency and frugality can be seen in clothes-drying. Few households nowadays possess a clothes-drying appliance, because few households consider such appliances to represent good value. The cost of the electricity or natural gas or propane consumed by the appliance is generally considered excessive in relation to the benefit provided. Consequently, almost everyone uses outdoor clotheslines whenever possible and indoor drying racks whenever necessary. Even wealthy people did likewise, as you know, not so many years before your time.

Geographical decentralization

I could offer you several additional examples of increased energy efficiency and frugality within the home. But as it turns out, most of our improvements in energy efficiency and frugality have occurred in activities that take place away from the home. Here too, just as in the case of household appliances, our improvements can be attributed primarily to new choices and new decisions rather than to new technology.

Probably the most important concrete result of our changeover to INF economics can be seen in the fact that we now conduct our economic affairs using only minimal amounts of transportation: minimal transportation of freight and minimal transportation of people. Now how, you may be wondering, can an economy function satisfactorily without a great deal of transportation activity? Well, we have not eliminated transportation entirely. Far from it. But your society was incredibly transportation-intensive. We have been able to save tremendous amounts of energy by simply removing the pro-transportation bias in the economy.

The key to minimizing transportation activity can be summed up in two words: geographical decentralization. We found it both necessary and desirable to decentralize almost every facet of the whole economy, just as Warren A. Johnson had imagined back in 1978 in his book *Muddling Toward Frugality*. Typically, that meant closing down one central unit (or at least reducing it in size considerably) and simultaneously opening up five or ten or fifty or a hundred much smaller units scattered throughout the territory previously served by the centralized unit.

At the retail level of grocery distribution, for example, decentralization typically involved (a) the elimination or major downsizing of one giant supermarket serving perhaps thousands of homes and (b) its replacement with a large number of much smaller grocery stores and butcher shops and so on, each serving perhaps a hundred homes or even fewer. The energy advantage of this decentralization is that most town-dwellers no longer need to use any kind of motor vehicle in order to transport their groceries from retail store to home. Instead, they use their own arms and legs, often supplemented by shopping buggies, wagons, bicycle trailers and so on, not to mention the major role played here by delivery boys and girls earning after-school pocket money.

Naturally this does not eliminate the need to transport groceries from their original source, which nowadays is usually not too far away, to the various retail food stores. But the energy efficiency involved in moving groceries by the truckload or cartload is incomparably greater than moving them a few bags at a time in private cars. And since people tend to buy groceries frequently, the energy savings attributable to decentralized grocery-shopping add up quite quickly.

We have also decentralized virtually all commercial food-processing activities. Consider dairies, for example. As in former days, dairies now exist not just in cities but also in villages and small towns throughout the milk-producing regions of the country. Milk, as you know, is fairly heavy. We are therefore reluctant to ship milk long distances from farm to dairy and then more long distances from dairy to consumers' homes. In most cases, that no longer happens.

INF economics has also helped restore the popularity of home cooking, home baking, home canning and root cellars.

Population decentralization and rural living

Slowly but surely, our whole population has been decentralizing itself as well. No longer do most of us live in a handful of heavily populated metropolitan areas.

Our farms today are much smaller and hence much more numerous than in your day. The small family farm is once again alive and well and prosperous. And the percentage of our population that is engaged in farming and homesteading has increased markedly. No one is compelled to "go back to the land", but under INF economics it is much easier to be a successful small-scale farmer than it was in your day.

Incidentally, on our farms and in our greenhouses, we no longer find it necessary or appropriate to bring in temporary migrant labourers from less wealthy southern countries. No farm owner would want to pay for all the costly transportation energy involved. Nor would we be happy with arrangements even remotely resembling the apartheid policies of the old South Africa.

The significant decrease in the size of the average farm has brought with it a number of changes in farming techniques. An increasingly important agricultural role is now being played by solar energy and solar-derived energy: human energy, animal energy, wind power, wood heat, small-scale hydroelectricity, biodiesel, solar-generated electricity and so on. At the same time, a smaller and smaller role is being played by fossil-fuel energy resources, both in the form of fuels and in connection with artificial fertilizers and other chemicals. The trick, our farmers and their families have found, is to steer nature gently and skillfully in the desired direction rather than to keep clubbing it over the head with heavy machinery and tankfuls of delinquent chemicals.

In the case of haying, for example, many of our farmers, just like a few in your day, use a team of draft horses to pull two complementary pieces of equipment, one behind the other. In front, immediately behind the horses, is a single-axled cart on which the farmer rides and on which is mounted a small internal-combustion engine. Behind that cart might be, say, a hay baler. The latter gets powered by the thrifty engine on the cart, but all the pulling work gets done by the team of horses. As for the engine's fuel, in most cases it is a solar-derived biofuel.

As we are pleased to acknowledge, we have received much inspiration from the many individual farmers and writers and members of religious communities who in your day, even though it was not popular to do so, expressed critical disapproval of the industrial approach to agriculture and who did their best to promote what they called agrarianism. We owe them all a debt of gratitude. Your economy made it so difficult to keep long-term values uppermost in mind, whereas ours helps to reinforce those values.

The combination of (a) large numbers of individual farms, (b) a much greater farming population and (c) decreased amounts of transportation activity has given our rural villages and small towns a tremendous shot in the arm. In fact, quite a few brand new rural communities now exist in places where in the recent past there was scarcely even a single house.

By reducing the distance between villages, we shorten the average distance from farm to village. That in turn generally shortens transportation distances both for outgoing farm products and for incoming supplies. It enables our rural population to travel shorter distances in the normal course of their day-to-day lives. And most of these observations apply to coastal fishing communities as well as to inland farming communities.

As for rural transportation itself, just about every imaginable energy-efficient method of conveyance gets used, depending upon the season, the circumstances and the individuals involved. People walk, bicycle, use other pedal-powered devices, ski, snowshoe, row, paddle, sail, use horses, use dogsleds, ride low-powered snowmobiles, ride or drive low-powered electric vehicles or low-powered vehicles with internal combustion engines, ride in low-powered motorboats, carpool, travel by low-powered bus and by low-powered or horse-pulled taxicab and so on.

In rural areas, just as in large cities, we now do most of our shopping as close to our homes as we possibly can. INF economics, by discouraging unnecessary transportation, effectively prevents our larger centres from attracting an excessive amount of retail business away from smaller communities.

There is more than retail commerce, however, behind the current prosperity of our villages and small towns. We have decentralized the actual making of things as well. Food-processing, furniture-making, metal-fabricating, clothes-making and many other similar productive activities are all likely to be taking place in hundreds of our small communities simultaneously.

Durability and Repairability

This is perhaps a good place to talk about durability and repairability. Durability, which includes ease of repair, has become an absolutely essential characteristic for both consumer goods and producer goods in our society. The high cost of any and all fossil-fuel energy ensures a very high demand for durability and a very low demand for disposability. In addition, the human energy necessary for repair work is untaxed and therefore attractively priced. Given these mutually reinforcing incentives, we repair almost everything. We recycle as well. But we like to do as much repairing and re-using as possible before recycling finally becomes necessary.

Consider, for example, "white goods": old refrigerators, stoves, washing machines and so on. Your tendency was to use such appliances for perhaps fifteen or twenty years and then treat them essentially as scrap metal. To us, that seems highly wasteful. We disassemble, clean, repair, refinish, reassemble and re-use our white goods over and over again. When, inevitably, certain components become worn or corroded beyond repair, we replace only the component in question, nothing else. We design repairability into almost everything we make.

It follows that one of the most important economic activities in all our communities, from big cities to rural towns and villages, consists of repair work. No one wants to ship a piece of machinery, a kitchen appliance, a musical instrument, a piece of furniture or a computer any farther for repairs than is absolutely necessary, because long-distance transportation tends to be expensive and in some cases a bit slow. Hence our repair facilities are just as decentralized as our production facilities.

Both for repair work and for production work, we make extensive use of hand tools, especially in connection with woodworking. We try to manufacture hand tools of the very highest quality, as we value real craftsmanship. Power tools and motorized equipment get used much more sparingly than in your day, and only in situations where the overall energy efficiency involved is high and where leaving the entire job to human energy and/or animal energy feels unreasonable.

Cities

This brings me to our cities. Just as in your day, our cities bustle with both economic and non-economic activity. They are therefore highly stimulating places in which to live. But a number of changes have occurred. Let me begin with urban transportation and, more particularly, with the urban transportation of people.

The first and most important difference here stems from the fact that most people nowadays live quite close to where they work. Commuting to work on foot, or by bicycle in decent weather, has thus become the norm. Public transportation comes second. At the rear come taxicabs and, in some cities, a few private cars.

In order to understand how all this could be possible, you need to keep in mind the decentralized nature of our economy. Small-scale decentralized industry has generally replaced large-scale centralized industry. Scattered here and there inside any typical city of ours, you will find countless small factories, workshops, repair shops and warehouses, not to mention local bakery shops, butcher shops, restaurants, retail stores, libraries, clinics, offices, theatres, schools, colleges, hotels, parks and religious institutions, all employing human energy on a significant scale; perhaps I should say "on a generous scale" in order to stress the welcome mat that INF economics lays out for human energy.

Our municipal by-laws no longer try to isolate residential areas from industrial and semi-industrial areas, except in special cases. Instead, the focus is primarily on dirt and noise and visual appearance, all three of which are fairly strictly regulated. Almost any productive activity, provided only that it is non-intrusive and environmentally benign, is permitted almost anywhere. So it is not really surprising that for most people the distance between home and work-place is short. By the same token, our cities generally consist of a number of local urban neighbourhoods; their boundaries are mostly unofficial, but their self-awareness and community spirit are recognized by everyone.

Concerning urban transportation, we believe that we have reclaimed our cities from the automobile. In fact, several of our larger cities have gone so far as to ban the private car altogether. Other cities are watching the experiment closely and may well follow suit. In the meantime, the general approach taken by most large municipalities has been to draw a clear line around the actual built-up part of the city. No motor vehicle of any kind is permitted to cross that line in either direction without stopping at a control gate. The purpose of these control gates is to enable the municipal government in question to collect a user fee from all persons choosing to bring a motor vehicle (car, truck, bus, motorcycle, etc.) inside the city limits.

This user fee is based on three factors: (a) the physical size of the vehicle, (b) the maximum quantity of tailpipe emissions, if any, that the vehicle is licensed to produce and (c) the number of kilometres (determined in one way or another) that the vehicle ends up being driven within the city limits.

In order to make such a system workable, we have had to add a few secondary features. In particular, all motor-vehicle fuels (biofuels as well as fossil fuels) sold inside the city are subject to a substantial surtax which, after being paid, qualifies as a credit towards payment of the user fee in question. The result is that our urban drivers pay an approximation of the user fee at each refueling within the city. Then when they leave the city, they stop at a control gate and either pay the balance owing or else receive back a cash refund for the amount of any overpayment. As for electric vehicles, their user fee generally includes a monthly charge for each month that they remain both (a) within the city limits and (b) licensed for use.

The revenue from all these user fees gets used for the payment of those municipal government expenditures that have been made necessary by the urban use of motor vehicles. Admittedly, a certain amount of urban motor-vehicle use can enhance urban living for everyone. But that is no reason, we feel, to reduce the user fees in question. All urban motor-vehicle operators are required to pay the full amount of those fees. Naturally, that raises the price of all motor-vehicle transportation services performed inside urban areas. Hardly anyone objects, however, because hardly anyone wants frugal households to have to subsidize extravagant households.

Now that we have had time to do some fine-tuning, everyone seems to have accepted the above constraints on the urban use of motor vehicles. Two conditions, however, have proved to be crucial. In each city, there must be a safe, clean and dependable system of public transportation, and there must also be a safe, clean and dependable fleet of taxicabs of one kind or another. But neither the one nor the other gets subsidized. That would be counter-productive, since our whole goal is to promote frugality and energy efficiency.

Nor do we want short trips to subsidize long trips. In large cities, accordingly, our transit fares increase with the distance travelled, even where the whole route lies within a single municipality.

Taxicab fares also merit a comment. Under our system, the cost of a motorized taxicab ride is by no means negligible. But neither is it a great deal higher than the overall cost of using one's own motor vehicle to cover the same distance. In most urban situations, the only extra cost is that of the taxi-driver's human energy: a very small—and untaxed—portion of the total energy being consumed. Consequently, most city-dwellers and many town-dwellers have voluntarily given up car ownership altogether and now use taxicabs on those occasions when they feel that travel by motor vehicle is appropriate.

As for bicycles, they have become extremely popular in both urban and rural areas. Because their use is so benign and so healthful, we have gone out of our way to make our society as bicycle-friendly as possible.

High-rise buildings do not have much appeal for us, either as apartment buildings or as office buildings. One problem is that in both cases large numbers of people have to be moved considerable distances to and from the high-rise building every day, not to mention all the energy-intensive elevator traffic. Such movement tends to be uneconomic. Another problem, which James Howard Kunstler identified in his book *Too Much Magic* back in your day, is that high-rise buildings are not at all easy to maintain and repair. And quite apart from economics, it is surely undesirable that any parents living in a country as spacious as Canada should have to raise their children in box-like apartments far removed from Mother Earth.

In the case of high-rise office buildings, another economic factor deserves comment as well. In your day, one of the most popular ways to obtain money was to profit, directly or indirectly, from early possession of information relating to financial markets and investment opportunities. Enormous numbers of person-hours were allocated, directly and indirectly, to the pursuit and appraisal of such information. Enormous quantities of office space were needed to sustain all this information-centred activity. So as not to miss out on the latest morsel of financial news or on related business, a great many people and a great many corporations and law firms were willing to pay substantial rental premiums in order to obtain centrally located office space in major metropolitan centres. High-rise office buildings met these needs perfectly. Such buildings even seemed to pay for themselves, despite consuming enormous amounts of energy and other resources, both in their construction and in their day-to-day operation, and despite producing within their four walls not so much as a shoelace.

Not so nowadays. Our economy functions quite differently from yours. Financial markets tend to be stable because the economy is so decentralized and so frugal. Potential profits attributable to the early possession of financial information tend to be tiny. High-rise buildings, especially those not designed to be very energy-efficient in the first place, have become very expensive to operate, and their high rental rates have proven very unattractive to prospective tenants. As a result, many of our inherited high-rise buildings have been torn down or reduced in size or have had their upper floors blocked off. New construction of such buildings is non-existent, regardless of the internal energy efficiency of their design.

Suburbs

Another change to our urban landscape can be seen in our suburbs, or rather in our lack of suburbs. Nowadays, the line between city and countryside is usually quite definite. Population densities are high in cities while being significantly lower in the countryside. Suburbs, being neither fish nor fowl, have pretty much disappeared. Depending on the situation, their place has been taken by (a) higher-density urban communities, (b) compact towns and villages serving the surrounding farms (c) re-established farms (d) re-established forests (e) re-established wilderness. A shortage of topsoil, however, has caused us a lot of headaches.

For the most part, these changes have simply resulted from the price structure provided by INF economics. The suburbs of your day produced too little and consumed too much. They were not able to survive, as suburbs, once we removed all subsidies in the transportation sector of the economy, including subsidies to the piped transportation of water and sewage.

Interurban Transportation

Two fundamental changes are worth noting in connection with interurban transportation. The first, already mentioned, is that in general we do not transport either people or freight nearly as often or nearly as far as in your day. That change alone has saved us large amounts of natural resources and has done the environment a great favour.

Secondly, we use a very different mix of transportation methods compared with your day. Whenever there is a transportation task to be carried out, we think first of all in terms of energy efficiency. In practical terms, this means that we are living once again in a golden age of railways, as regards both freight and passengers. The energy advantages of rail transportation over both air transportation and long-distance highway transportation are in most cases overwhelming. Transportation by ship and by barge and by ferry, where possible, can also be attractive. Otherwise, all eyes are on railways.

It follows that our transportation scene looks quite different from yours. The glory of the downtown railway station has been restored. Long-distance travel by car is rare. Eighteen-wheel (and larger) transport trucks are seldom seen. Most of the land once occupied by superhighways has been allocated to other uses. Many large highway bridges are no longer in service.

How fast do our trains actually travel? Nowhere near as fast as the high-speed passenger trains of your day in such countries as France and Japan. Our attention is focused on optimum speeds rather than on maximum speeds. Our technological efforts are aimed at raising the levels of safety, comfort and energy efficiency. High speed is energy-intensive and therefore expensive. In our user-pay economy, we have found that few consumers consider high speed to be worth the money it costs. Something of a parallel existed years ago in the unwillingness on the part of the vast majority of air travellers to pay the hefty premiums required for passage aboard the supersonic Concorde aircraft.

As for commercial air flights, whether for passengers or for freight, our high prices for large concentrations of energy have essentially rendered such flights unmarketable. Admittedly, some diehards and some airship enthusiasts still have stars in their eyes! For the most part, however, we are content to do our long-distance travelling more slowly and less frequently than in your day and without leaving the surface of the Earth.

Back in 1974, as you may know, Ivan Illich wrote a little gem of a book titled *Energy and Equity* in which he explained an additional reason for not moving human beings around at high speeds. He summed up his point as follows: "High speed is the critical factor which makes transportation socially destructive." We agree, based on our study of the past.

Synthetic Chemicals and Environmental Protection

If I may turn now to synthetic chemicals, our annual consumption of such chemicals represents only a tiny fraction of the amount considered normal in your day. In agriculture, for example, we rarely use synthetic pesticides any more. In industry generally and in the home as well, we have in a great many cases been able to solve our day-to-day problems without much help from synthetic chemicals. Here, as elsewhere, our successes can be traced mostly to our technological choices rather than to any new technological inventions or discoveries.

The key factor in steering us away from synthetic chemicals has once again been price. Your petrochemical industry relied on extremely low prices for fossil-fuel energy resources, using those resources both for feedstock and for fuel. Given those low prices for inputs, the various chemicals you produced had little difficulty in snatching work away from human beings and in establishing very large niches for themselves. Not so with us. Because of our very different taxation system, synthetic chemicals are in most situations too expensive to be competitive. As a result, the marketplace itself has rejected them.

Not that we leave everything to the marketplace. If any particular chemical, whether synthetic or otherwise, appears to pose a significant threat to human health and safety, then we do not hesitate to use regulations and/or prohibitions to overrule the marketplace. In theory, your generation did the same. But our task is much easier than yours was because we have not saddled ourselves with the enormous economic pressures—caused by your overall price structure—that in your day made it so difficult to keep the chemical lid firmly in place.

More generally, our whole effort at environmental protection has met with fewer obstacles than in your day. The main reason, I think, is that our economy makes do with far smaller amounts of raw resources than yours did. Consequently, neither our "sources" (of raw resources) nor our "sinks" (into which we deposit, in one way or another, the various solid, liquid and gaseous waste products of our economy) are associated with the enormous scale of environmental destruction and degradation that you folks have found to be so challenging.

Conclusion

I should probably stop at this point. You can no doubt imagine many additional details of our lives here, based on what I have already described. The essence of it all is that we have installed human energy in the royal apartments, we have demoted fossil-fuel energy to the scullery kitchen and we have thrown nuclear power out the palace window!

Let me end by assuring you that, despite all my talk about energy efficiency, we regard sound economics as being merely a means to an end. Ultimately, as I am sure you would agree, life is about trying to leave the world a better place than one found it. INF economics, in my view, has helped us keep that in mind. It has done so primarily by providing everyone with plenty of meaningful and appropriately remunerated work opportunities. Moreover, it has permitted us to invest large numbers of man-hours and woman-hours, as distinct from material goods, in the upbringing, education, training, supervision, advising and encouraging of our children and our young people. The task of passing the torch never ends, and it is surely one of the most important tasks facing any society in any era.

Here in the society and era in which I am living, we do not have all the answers by any means. Nor do we always resist the temptation to bury our head in the sand, just like anybody else, even when we know what needs to be done. Human nature does not seem to change much from one generation to the next!

We all wish you the very best.

With much affection,

A direct descendant

P.S. Don't be afraid of taking the first step. History has given you a unique opportunity. As Thomas Berry wrote in 1999, "For the success or failure of any historical age is the extent to which those living at that time have fulfilled the special role that history has imposed on them." And Naomi Klein relayed the following question at the end of her 2014 book *This Changes Everything*: "History knocked on your door, did you answer?" Cheers!

Part 2: Putting It All Together

4

Making the Tax Progressive

"Our recent experience shows that the question of progressive vs. regressive taxation is not one that can be postponed to some more propitious time, nor can it be safely separated from other social concerns."

George P. Brockway (1995)

"Thus proponents of environmental taxes (and subsidy cuts) ignore the issue of regressivity at their peril."

David Malin Roodman (1998)

This chapter sets out a method by which the proposed new tax on fossil-fuel energy can be converted from a regressive tax into a progressive one. A regressive tax is defined as a tax that imposes a relatively heavier burden on the poor than on the wealthy. A progressive tax does the opposite.

If no further provisions were built into Intelligent National Frugality (INF) economics, one could protest that people with low incomes would suffer severe hardship from the new tax on fossil-fuel energy resources. Low-income city-dwellers in particular might not be able to afford even the most basic quantities of energy necessary to keep their homes provided with minimum amounts of heat, light, hot water, cooking energy and refrigeration energy.

The problem is that the tax as outlined is not progressive. Low-income earners would pay the same rate of tax on natural gas, for example, as would high-income earners. Contrast this with today's personal income tax where (at least in theory) high-income earners pay not just more taxes in absolute dollars, but also a higher rate of tax than do low-income earners. Moreover, people with very low incomes do not at present have to pay any income tax at all.

It makes sense, therefore, to incorporate into INF economics a fairly simple feature that will have the effect of converting the new tax into a progressive tax. I will identify this feature in a moment, but first let us look at the matter in conceptual rather than practical terms.

Conceptual progressivity

Conceptually speaking, we want to convert the new tax into a progressive tax by setting up a rationing system. What we want to ration is not fossil-fuel energy itself, but rather a tax exemption. We want to grant each and every person in the whole country the right to purchase a limited quantity of fossil-fuel energy on a tax-free basis. The limit must be the same for everyone, rich and poor alike. And once a person has purchased his or her limit, he or she must then pay the full price for any and all additional purchases of fossil-fuel energy. In short, the first bit must be tax-free for everyone, while any extra must be fully taxed for everyone.

Note that no limits at all are being proposed for the quantities of fully taxed fossil-fuel energy that any individual person might wish to purchase. It is only the tax exemption, not the energy itself, that is being rationed. To continue with this conceptualization, imagine a system in which everyone received a ration card entitling the card-owner to a tax exemption on the purchase of up to, say, twenty units of fossil-fuel energy every month. (Since this is just an example, there is no need to specify how large or how small such units might be.) Forget, for the time being, the bookkeeping complications. Each card-owner would be entitled to purchase this tax-free energy in whatever form he or she wished. It could be in the form of gasoline or natural gas or furnace oil or fossil-fuel-derived electricity. Or it could be in the form of some piece of merchandise, such as an engine or a boat or a package of roofing materials, whose manufacture and delivery had consumed the rationed amount of fossil-fuel energy. Or it could be any combination of the above.

Energy losses are worth a comment here. In all manufacturing, refining, shipping and other such processes, including the generation and transmission of electricity, there are unavoidable energy losses. Hence a general ration of twenty units of tax-free fossil-fuel energy translates into a specific ration of significantly less than twenty units of gasoline energy at the local service station. Similarly, it translates into significantly less than twenty units of fossil-fuel-derived electrical energy passing through one's household electricity meter. In other words, all such energy losses would use up a portion of the tax-free ration that we are discussing. If this were not the case, i.e. if the card-owner's ration did not have to cover the relevant energy losses as well as the energy actually being received, then the card-owner would have no incentive to shy away from purchasing items that for one reason or another happened to embody large energy losses. Such an incentive is needed, however, in order for the tax to do its job properly.

There is nothing new about this idea of energy consumers having to cover the costs of relevant energy losses. In Ontario, for example, the monthly invoices issued by Hydro One to residential electricity customers state, albeit in tiny print, "The delivery charge also includes costs relating to electricity lost through distributing electricity to your home or business ... When electricity is delivered over a power line, it is normal for a small amount of power to be consumed or lost as heat.

Equipment, such as wires and transformers, consumes power before it gets to your home or business."

Let me return to the ration cards. Once the card-owner had purchased twenty units of tax-free fossil-fuel energy, all subsequent purchases of fossil-fuel energy during the month would be fully taxed.

Conceptually speaking, a ration-card system of that kind is exactly what we need. It would convert the fossil-fuel energy tax from a regressive tax into a progressive tax. More precisely, it would cause the effective tax rate to depend upon one's total monthly purchases of fossil-fuel energy. For those who buy no more than twenty units of such energy per month, the effective tax rate would be zero. For those who buy forty units per month, the effective tax rate would rise to 50% of the nominal rate. For those who buy one hundred units per month, the figure would climb to 80%. And for those who splurge and buy one thousand units per month, the figure would reach 98% of the nominal rate. The more fossil-fuel energy anyone buys, the greater would be his or her average cost for each energy unit bought.

Practical progressivity

Now how can this abstract conception be translated into a practical and workable system? It turns out that neither the ration cards nor the bookkeeping entries are needed. All that is necessary is for the government to make a regular monthly (or quarterly) payment to—or, in the case of children, on behalf of—every individual resident in the whole country.

To continue with our previous example, suppose that the government wishes to ration a tax exemption on twenty units of fossil-fuel energy every month. In that case, the government need only issue everyone with a monthly cheque (or a direct-deposit equivalent) for exactly twenty times the amount of the tax levied on a single unit of energy. Nothing else is necessary. The effect is exactly the same as the intended effect of the conceptual rationing system involving ration cards and bookkeeping entries. Moreover, no one has to worry about trying to estimate the indirect fossil-fuel energy input in every screwdriver, every milkshake, every newspaper and every visit to the dentist. Once such a rationing system has been set up, many people will no doubt direct the government to remit their monthly ration payment directly to one or other of their major energy suppliers: a furnace-oil company, a natural gas company, an electrical utility. In that case, the energy supplier will be required to credit the full amount of the ration payment to the account of the customer concerned. But anyone who so chooses will be entitled to receive his or her ration payment directly.

Admittedly, a government bureaucracy will be needed to administer this rationing system. But, not having to concern itself with anyone's annual income, it will be a far smaller bureaucracy than the one we have at present. And anyone preferring not to participate in this rationing system will be at liberty to have his or her name expunged from the government's list of ration payees!

How large or how small should these ration payments be? If made monthly, should they be equal to the average tax payable on 1 litre of furnace oil? On 10 litres? On 100 litres? This is essentially a political question, and the answer will have to be worked out by the political process.

Once it has been decided upon, the size of the standard ration payment will determine the degree of progressivity of the tax. If, for example, the monthly ration payment is so small that it is only equal to the average tax on 1 litre of furnace oil, then the overall effect of the tax is not going to be very progressive at all. But if the payment is large enough to equal the tax on 100 litres of furnace oil, then the progressivity will be quite significant.

It may look as though there are only two factors involved here: the size of the ration payment and the degree of progressivity of the tax. But, as a moment's reflection makes clear, a third factor is intimately involved as well: the rate of the tax. Indeed, if the size of the ration payment is increased, then the rate of the tax will almost certainly have to be raised as well. This is because a portion of the gross revenue generated by the tax will have to be used to fund all the ration payments. The larger the sum total of all the ration payments, the larger the gross revenue needed from the tax. Otherwise, the government will not have enough net revenue to finance its normal activities.

There are thus three little children all holding hands and all dancing around the Maypole together: (a) the rate of the tax, (b) the size of the ration payment and (c) the degree of progressivity involved. Getting all three coordinated in a manner that is both socially and economically satisfactory will definitely involve the political process.

What about the situation where some individuals live so frugally that they actually make a profit on part or all of each ration payment they receive? Anyone who regularly receives a monthly ration cheque for $50, for example, and who during one particular month spends only $40 in total on the fossil-fuel energy tax will earn a profit of $10 in that month. But there is nothing wrong with that. Such profit can be considered as a cash reward from society to those individuals who have made singular achievements in the field of personal frugality. INF economics positively welcomes intelligent personal frugality.

Should ration payments be made to corporations, schools, hospitals, charities or any other collective bodies? Not at all. Only living flesh-and-blood human beings need receive such payments, since only living human beings would be at risk of suffering in the absence of such payments.

How long should these ration payments continue? No decision on that point needs to be made in advance. Once INF economics is fully established, any future government will be able to increase or decrease or maintain unchanged, as it wishes, the amount of the ration payment. The progressivity of the tax will always be subject to deliberate change via the democratic process, just as is the progressivity of our present personal income tax.

Leftist versus rightist leanings

Not everyone is likely to be in agreement concerning the optimal size of these ration payments. People having leftist or socialist leanings will probably tend to want the payments to be quite large. And people with rightist or laissez-faire leanings will probably tend to want the opposite. In any case, the democratic political process ought to be able to deal with non-unanimity on this matter, just as it does on so many other matters.

Similar proposals, different names

Other writers too, using terms such as "ecobonus" or "Citizen's Income" or "Basic Income" or "a tax on energy that would be rebated on a per capita basis", have put forward proposals similar to, if not identical to, the one described in this chapter. James Robertson, for example, advocates that for progressivity reasons the government pay a Citizens' Income "to every man, woman and child as a right of citizenship. It will be age-related, with more for adults than children and more for elderly people than working-age adults." Robertson then adds further details: "There will be supplements for disability, housing benefits, and other exceptional circumstances. Otherwise [Citizen's Income] will replace all existing benefits and tax allowances ... [and its amount] will be unaffected by [the recipients'] income or wealth, their work status, gender or marital status." Clearly, there are certain small differences between Robertson's proposal and my own. But my basic point here is that many other people have incorporated a progressivity component into their energy-tax (or similar) proposals.

A small conceptual difference

Finally, I should acknowledge that there is a small conceptual difference between (a) the progressivity of a normal personal income tax and (b) the progressivity of the taxation system described in this chapter. In the former case, the progressivity depends only upon the income itself and not upon how the income gets spent. But in the case of the INF system, the progressivity depends upon how the wealthy actually spend their money. Suppose, for example, that a certain wealthy individual took pains to consume, directly and indirectly, no more than a very small quantity of fossil-fuel energy every month. In that specific case, the INF tax, even when combined with the system of ration payments described above, could not be said to be functioning as a progressive tax.

As far as I can see, however, no practical problems arise from this conceptual difference between the two versions of progressivity. Suppose that a certain number of wealthy people did choose to forgo what wealthy people in the modern world almost never do forgo (unless they give their money away and become "formerly wealthy" people), namely the multifarious pleasures associated with energy extravagance. Such abstinence is in fact socially desirable and to be encouraged. It does not constitute a problem at all. Two other problems, which might conceivably accompany such abstinence, can be dealt with on their own, as follows: If government revenue happens to fall short of reasonable expectations, then the rate of the INF tax can be raised. And if our inequality of wealth is judged to be excessive, then a wealth tax and/or an inheritance tax and/or a tax on very high personal income can be imposed or, if already in existence, increased in size. Otherwise, the progressivity problem can simply be dealt with as set out in this chapter.

5

Exports, Imports and Free Trade

"Public understanding of the narrowness of free-trade agreements and the absurdities they encourage goes beyond that of politicians and elites."

Hazel Henderson (1996)

"But free traders say we should become less self-sufficient and more globally integrated as part of the overriding quest to consume ever more. That is the worst advice I can think of."

Herman E. Daly (1996)

Inappropriate foreign trade

Intelligent National Frugality (INF) does not have a bias either in favour of or against foreign trade as such. It does have certain biases, however, some of which are highly relevant to foreign trade. In particular, it is biased against any extravagant or inefficient or otherwise inappropriate use of non-renewable energy. It is also biased against unreasonable resource depletion. On the positive side, it strives to promote full employment, economic justice, community strength, environmental responsibility and a reasonable degree of national, regional and local self-sufficiency.

There would therefore be no good reason for any country, once having adopted INF economics, to stand by and allow inappropriate kinds of foreign trade to curtail potential INF achievements. Put more simply, there is no point in tying a knot with the left hand, only to have it untied with the right. Or as

Wendell Berry asks, "How can any nation or region justify the destruction of a local productive capacity for the sake of foreign trade?"

Under INF economics, therefore, it seems virtually certain that Canada (and any other country similarly inclined) will want to withdraw from any and all general free-trade agreements with other countries. Such a withdrawal will permit us to march to our own drummer even if no other country in the world has yet adopted INF economics. It will permit us to restrict any of our imports as well as any of our exports, as we see fit. At the same time, it will still allow us to carry on mutually beneficial trade with any country willing to accept our terms.

Accordingly, throughout this chapter and in fact throughout this book, I will assume that where exports and imports are concerned our hands are in no way tied.

Exports

Let us begin with exports. Under no circumstances should a rebate of the fossil-fuel energy tax be granted for exports: not for exports of fossil-fuel energy resources themselves and not for exports of goods whose manufacture and transportation may have involved the consumption of fossil-fuel energy resources.

Here is the reason. Three of the main goals of the fossil-fuel energy tax are (a) to minimize the amount of environmental degradation attributable to the consumption of our domestic fossil-fuel energy resources, (b) to continuously slow down the depletion of all our domestic non-renewable resources and (c) to prevent entirely the depletion of any of our domestic renewable resources. All three of those goals would tend to be compromised if tax rebates were granted for exports. Not only should tax rebates for exports be ruled out, but we might even choose to impose export taxes on certain commodities. The reasoning in such a case would be that the commodities in question were worth more to Canadian society in the long run—including to future generations of Canadians—than the price for which their current owners were willing to export them. Scrap steel constitutes a possible example. So do fossil fuels themselves.

We might go further still. We might actually ban the export of certain scarce resources such as one or more of the fossil fuels. Such a ban can be thought of as an investment: We would be forgoing present consumption of wealth in order to obtain, for ourselves or for our descendants, a future benefit that would otherwise be unavailable. Admittedly, we have all been brought up to feel that mere abstention cannot possibly by itself constitute an investment. Perhaps only people with limited quantities of food or water like the lost survivors of a desert plane-crash can really understand how much importance to place on the future benefits of present abstention.

Imports

As for imports, clearly the new tax must be imposed on all imported fossil-fuel energy resources, whether unrefined or refined. Otherwise, most of the intended benefits of INF economics would be lost, and in addition we would probably saddle ourselves with a severe balance-of-payments deficit vis-à-vis the rest of the world because of all the inexpensive fossil-fuel energy resources we would end up importing from non-INF countries.

Applying the tax to imports raises the question of double taxation. What if Canada wished to import some crude oil, for example, from a country that had already imposed a hefty tax on that same oil? But there is nothing in principle wrong with such double taxation. If, after both countries' taxes had been factored into the final price, there was still a market in Canada for this crude oil, then the transaction would proceed. If not, then it would be stopped short. Neither case would pose any particular problem.

Where the question of imports becomes less clear-cut is in the area of indirect energy consumption. For example, a considerable consumption of nuclear and/or fossil-fuel energy may have taken place outside Canada in connection with the manufacture and transportation of certain foreign goods. How should INF economics deal with the importation of such goods into Canada?

One approach would be to allow such imports to enter Canada on a tax-free basis. Canada, it could be argued, would thus benefit from being able to use and enjoy the goods, while the exporting country would bear the cost in two important areas: environmental degradation and diminished supplies of energy resources.

But that argument is not exactly uplifting. Moreover, it overlooks two important negatives: (a) Canada would lose jobs and (b) the Canadian government would lose tax revenue. It also overlooks the fact that environmental degradation often disregards international boundaries.

A wiser approach to the importation of foreign manufactured goods would surely be to impose the INF energy tax—in the form of an import tariff—on the estimated amount of nuclear and/or fossil-fuel energy consumed in the manufacture and transportation of the item being imported. In order to be able to assess such tariffs competently, our customs officials would need to possess considerable expertise. But employing expertise for the purpose of keeping the national economy on an even keel has always been considered a sound investment.

There is one situation where it might make sense to bend the rules. Western Canada, for example, might wish to export a certain quantity of coal to western U.S.A., while eastern Canada might wish to import an equivalent amount of coal from eastern U.S.A. Such an exchange could obviate the need for transcontinental coal haulage in both countries. For that reason, it might make sense under those specific circumstances for each of the two countries to allow the matched amount of incoming coal to enter on a tax-free basis.

Money versus real resources

This may be a suitable place to assert that Intelligent National Frugality tries never to confuse money with real resources. Money is worth little or nothing in the absence of real resources. But real resources, as every would-be Robinson Crusoe knows, can have enormous economic value even in the absence of money. Admittedly, money and wealth do seem to be virtually the same thing from the point of view of the individual person living in modern society. But for society as a whole, it is vital to remember that converting a tree into cash involves not just the gain of the cash but also the loss of the tree.

Much of the modern world at present seems fiercely intent on converting its real resources into cash. It is easy to go too far too fast, as the collapse of our east-coast cod fishery has unfortunately shown.

Under conventional economics, excessive international trade can easily lead to situations where, for example, one country has depleted its timber reserves, a second has lost much of its topsoil, a third has poisoned its inland waters and a fourth has prematurely run out of oil. Surely all four countries here have ended up as losers regardless of what the account books and bank statements may say.

International transportation

International trade in goods necessarily involves international transportation. At present, such transportation is heavily subsidized, especially when looked at from the point of view of INF economics. In particular, the fossil-fuel energy used to transport international shipments by plane, by ship, by train and by truck is heavily subsidized. If adopted worldwide, INF economics would remove all those subsidies. As a consequence, many of the absurdities of modern international trade and transportation (such as the one described at the beginning of this book) would be eliminated. In the meantime, INF economics will allow Canada to resist on its own the current allure of subsidized international transportation.

Herman Daly describes one transportation absurdity humorously: "For example, the United States imports Danish butter cookies, and the Danes import U.S. butter cookies. Somewhere on or above the North Atlantic the cookies pass each other. Surely the gains from trading such similar products cannot be large. But regardless of their size, could not these gains be had more efficiently simply by exchanging recipes?"

Most of us in our daily lives have little opportunity to look in on all the various kinds and amounts of cargo that conventional economics finds it perfectly normal and reasonable to load into jet aircraft and then transport all over the world. But in his essay "Flight", Barry Lopez has given us a revealing portrait of the modern world's air-freight industry. Among the air-cargo items mentioned in that essay are horses, cattle, luxury cars, 70,000 pounds of gold, a killer whale in a tank, personal computers, clothing, a racing yacht, a tropical-hardwood bowling alley, 17 cartons of basketballs and 5,898 pounds of sunglasses. "What planes fly, generally," observes Lopez, "is what people imagine they want. Right now."

The word "people" in that last quotation should not be taken as being synonymous with "consumers". Producers too, in making perfectly rational business decisions under conventional economics, can involve themselves in almost unimaginable transportation absurdities, as I have already pointed out. In another example, *The Globe and Mail* reported on June 29, 2002 that a Japanese auto maker was planning to airlift to North America from Japan "about 2,000 tons of high-quality steel ... in 20 shipments, with more to come." Since even under conventional economics that particular kind of transportation absurdity is uncommon, the newspaper added an explanatory sentence: "Airlifting of steel is a highly expensive proposition that is resorted to usually when companies are desperate to keep their assembly lines operating." The desperation in this particular case stemmed from changes to U.S. tariff policies!

These examples call to mind Garrett Hardin's classic essay "The Tragedy of the Commons" wherein a few perfectly rational owners of cattle, perceiving that it is in their own narrow self-interest to do so, introduce one or two extra cattle into the village commons. In the short term, the extra cattle graze and put on weight and increase in value for their owners. In the longer term, overgrazing destroys the commons and hence everyone loses. In similar fashion, modern transportation absurdities earn money in the short term for the various participants, but in the long term we all lose.

Economic competition

What about the argument that free trade promotes economic competition amongst a large number of independent producers and thus keeps consumer prices as low as possible? As it turns out, INF economics not only accepts the desirability of maintaining large numbers of independent producers, it actually goes much farther than conventional economics does in translating that theory into reality. The key point here is that INF economics will create a highly decentralized domestic economy. Large numbers of small domestic producers will enter almost every reasonable niche in the whole domestic economy. To whatever the desired extent, therefore, domestic economic competition will be healthy and widespread, even without much of a spur from foreign producers.

Domestic economic competition actually possesses two major advantages over international economic competition, quite apart from the question of transportation absurdities. One advantage is that domestic employment tends to remain strong, especially under a system (such as INF economics) that favours human energy in the first place. The other is that international economic competition can easily induce us to accept far lower social and environmental standards inside our own country than we would otherwise tolerate.

6

Lower Levels of Government

"In each province, the legislature may make laws in relation to the raising of money by any mode or system of taxation in respect of (a) non-renewable natural resources and forestry resources in the province and the primary production therefrom, ... but such laws may not authorize or provide for taxation that differentiates between production exported to another part of Canada and production not exported from the province."

Section 92A(4) of the Constitution Act, 1867

This chapter looks briefly at how Intelligent National Frugality (INF) economics might affect lower levels of government (provincial, territorial and municipal) in Canada.

Municipalities

Technically speaking, municipal taxes (i.e. so-called "property taxes") could be completely abolished under INF economics. For that matter, they could also be completely abolished under conventional economics. Without tax revenue of their own, municipalities would have to receive their general revenue from a higher level of government.

But there are two good reasons for retaining some form of municipal taxation based on land ownership. One reason is to provide each municipality with at least a limited amount of independence and autonomy. The other reason is to impose an annual financial charge on landowners for the ongoing benefits that they receive by virtue of their ownership of the land in question.

In connection with the second reason, the question arises as to whether municipal taxes should be imposed on both land and buildings, or whether such taxes should be restricted to land only. At present in Canada and in many other countries, municipalities impose taxes on both land and buildings. But a strong argument, one that I too support, has been put forward in favour of restricting such taxes to land alone. Why, for example, should an annual tax be payable on a kitchen and a bathroom in cases where those facilities are located inside a building but not in cases where they are located inside a travel trailer or a motor home or a yacht? Such inconsistency and arbitrariness tend to bring the whole concept of taxation into disrepute.

Moreover, there is a very important specific reason for not taxing buildings at all. As David Malin Roodman has written, "Taxes on buildings are taxes on work and entrepreneurship. They reduce the supply of buildings, raising home and office prices and penalizing urban development and redevelopment." I would add that any tax on work and entrepreneurship falls into the category of a tax on human energy and is therefore objectionable for all the reasons set out in this book. In particular, why penalize someone financially, year in and year out, just because the person has made an effort to improve the value of his or her home? Why not let the fossil-fuel energy tax of INF economics nudge all homeowners—and all owners and would-be owners of other buildings—in the direction of durability, timely repair, energy efficiency and frugality?

The fact that our real-estate law considers buildings to be part of the land on which they are located does not mean that our municipal taxation law is compelled to follow suit. Our legal system is perfectly capable of making such distinctions.

Under INF economics, then, municipalities will probably no longer impose taxes on buildings at all. A municipal tax on the market value of all privately owned land, however, will no doubt be retained. In other words, all property assessments for municipal taxation purposes will have to disregard the value of any and all buildings located on the land in question. On the other hand, the new energy tax will be imposed on all the fossil-fuel energy consumed by every single building in the whole country.

As suggested in Chapter 3, an excellent case can also be made for the imposition of an additional municipal tax or fee: a tax on the use of motor vehicles inside urban municipalities. Precisely because of the high population density of cities and large towns, space therein is at a premium. So is clean fresh air. And so is reasonable quietness. Anyone operating a motor vehicle inside the limits of an urban municipality cannot help but impose significant constraints on the freedom of others: the freedom of others (a) to have plenty of elbowroom in public outdoor spaces, (b) to make safe use of urban streets, (c) to breathe clean fresh air and (d) to experience a reasonably quiet environment. Incidentally, even electric vehicles (whether powered by batteries or fuel cells), even ones equipped with regenerative braking, sully urban air somewhat as their tire treads and brake linings slowly wear down and turn to dust.

Regulations can limit such constraints. But they cannot eliminate them entirely, unless all motor vehicles are totally banned from the municipality in question. It therefore seems reasonable that any municipal expenditures occasioned by the local use of motor vehicles should be paid for, on a user-pay basis, by the vehicle operators. After all, we are talking here about the exercise of a privilege, the privilege of operating a motor vehicle inside a particular municipality. Those who are granted privileges ought surely to pay the costs occasioned by their exercising of those privileges. And if those privileges are being exercised by a producer, as distinct from a consumer, then the producer can be expected to incorporate the cost of those privileges into the price of whatever product or service he or she is providing.

Under INF economics, the combination of (a) municipal land taxes and (b) municipal user-pay fees levied on motor-vehicle use and (c) municipal regulations and bylaws will enable each individual municipality to shape its own destiny to a very considerable degree.

Municipal water

Potable water of course surpasses even fossil-fuel energy as a basic human need.

In the case of farming families and other low-density rural dwellers, each household can reasonably be expected to be responsible for its own water supply (typically from wells or springs) under normal circumstances. But in cities, towns and the majority of compact villages, municipal water systems are essential. Such systems can be operated either by private enterprise or by municipalities themselves. For the sake of brevity, I will discuss the latter situation only.

Just as with fossil-fuel energy, the question arises as to how best to set up a pricing system for municipally supplied water. In the absence of subsidies, such water will tend to be somewhat costly under INF economics, at least at first, because in most cases significant quantities of fossil-fuel energy will probably be needed for water pumping and water purification. Not everyone is likely to be able to afford reasonable amounts of such water.

A suggested solution for this problem would borrow some ideas from the rationing system described in Chapter 4. But instead of rationing a tax exemption, each municipality could simply ration a monthly allotment of subsidized water to every household located within the municipality. Every household could thus purchase (a) a limited quantity of low-priced water every month and (b) an unlimited quantity (or perhaps an additional limited quantity) of full-priced water every month. Businesses and other institutions would all pay full price for any and all municipal water they consumed.

Subject to whatever standards and regulations might be imposed by higher levels of government, each municipality could work out for itself the details of its subsidy system. Those details would include the setting of the two different water prices: the subsidized price and the non-subsidized price. They would also deal with the question of whether the subsidized quantity of water for a large household should or should not be greater than for a small household.

Each municipality would also be able to decide for itself where to obtain the funds with which to cover the water subsidy. They could be obtained from the municipality's general revenue or from its revenue from the distribution of unsubsidized water or partly from the one and partly from the other.

Apart from this section of this chapter, water does not receive a great deal of explicit attention in this book. But its importance— and the heedlessness with which conventional economics contaminates and wastes it—should never be overlooked.

The provinces

With regard to our provincial governments, two different approaches deserve attention. Economically, the two seem equally valid. But politically, the differences are significant.

In the simpler of the two approaches, the provincial governments do not levy any general taxes of their own at all. Instead, they rely on the federal government. The latter, having levied the fossil-fuel energy tax on all fossil-fuel energy resources entering the national economy, then remits to the provincial governments a fixed percentage of the total revenue from that tax. Each province receives a share that is proportional to its population. As a result, all provinces are treated alike.

The second approach, unlike the one just described, takes into account the identity of the province of origin in all cases where domestic fossil-fuel energy resources are involved. In doing so, this second approach permits and encourages double taxation.

The crude-oil reserves located within the province of Alberta can be taken as an example. Under this second approach, the Alberta government would impose its own fossil-fuel energy tax—over and above the tax levied by the federal government—on the energy content of every barrel of crude oil pumped up from Alberta soil. Alberta could set the rate of its own tax at any level it wished. As a result, Alberta crude oil would be subject to double taxation, just like imported crude oil. But as I suggested in the previous chapter, there is nothing inherently wrong with double taxation.

The chief attraction of the second approach is political. No one would have to worry about how to persuade provinces such as Alberta to relinquish something they would almost certainly prefer to retain: their right to tax their own fossil-fuel energy resources. (Whether such a tax is called a tax or a royalty does not matter in this context.)

Those who live in energy-poor provinces might object to this double taxation. They could claim, quite correctly, that its likely effect would be to transfer money from energy-poor provinces to energy-rich ones. But is that unfair when fossil-fuel energy is being transferred in the opposite direction?

The actual impact on Canadian society of any provincial fossil-fuel energy tax would depend upon the size of the latter. A small provincial tax could no doubt be taken by everyone in stride. But a large provincial tax would have major effects.

Suppose that Alberta, for example, were the only Canadian province in possession of reserves of crude oil. Suppose further that Alberta decided to impose a large energy tax—over and above the federal energy tax—on its own oil. Double taxation would replace single taxation. For all Canadians, Albertans as well as non-Albertans, the price of Alberta oil would increase by a large amount.

Depending upon what was happening in the rest of the world, the effect of this large Alberta tax might be to price Alberta oil right out of the Canadian market. If so, all of Alberta's oil would end up staying in the ground until circumstances changed. The whole of Canada, including Alberta, would in the meantime have to import from other countries 100% of whatever quantity of oil it wished to consume. Under such circumstances, the Alberta tax on its own crude oil would generate no revenue at all.

The above does not seem too likely a scenario. So let us suppose that the Alberta tax on its own oil was large but not prohibitively so. In that case, all oil supplied to the Canadian market might well come from Alberta and thus be subject to hefty double taxation: a large tax payable to Ottawa and another large tax payable to the Alberta government.

What would Alberta do with all its extra revenue? It would already be receiving a presumably adequate amount of revenue from the federal government, just as would all the other provincial governments. Would Alberta use the extra revenue to make cash payments to its own residents? Or to improve the level of provincial government services in Alberta? Or to invest in some heritage fund intended to benefit all Albertans after the province's oil reserves have been depleted? Or some combination of the above? And depending on the choice made, would large numbers of Canadians from other provinces be tempted to move to Alberta to share in the enjoyment of all this extra cash?

Since these questions cannot be answered in advance, all one can say right now is that under INF economics the shape of Canadian society as a whole may possibly be affected significantly by the taxation policies of certain provincial governments.

In any case, the major benefit of INF economics to all our provinces and territories will not involve interprovincial or interregional transfers of money. Rather it will be found in the general decentralization of economic activity throughout the whole country. Every region of Canada, to the extent that it has assets of some kind to work with, can be expected to become resourceful and productive, with plenty of work available for all who live there. What will get transferred to those provinces and territories that have hitherto been disadvantaged will not primarily be either money or fossil fuels but rather the opportunity and the incentive to become a local creator of wealth. Local and regional self-sufficiency, while not likely to become absolute, will develop to a much higher degree than anything even imaginable under conventional economics.

I should also mention mineral resources, i.e. non-energy mineral resources such as iron ore and copper ore. Under INF economics, Canadian mineral resources will still come primarily under provincial jurisdiction. Each provincial government will therefore still be able to offer its mineral resources for sale at whatever price it chooses. If substantial, the offered selling price (whether called a tax or a royalty) will likely have the effect (a) of further restricting the market for the resource in question and hence (b) of leaving even more of the resource available for future generations. In short, any provincial minerals tax will tend to reinforce the frugality that is already built into INF economics.

The territories

In principle, the territorial governments of the north can be treated under INF economics in much the same manner as the provincial governments of the south. Some or all of the territories may even find their evolution into full-fledged provinces hastened, given the degree of economic decentralization to be expected in all regions. In the meantime, the federal government will have to decide how closely its per capita revenue transfers to the territorial governments should match those allocated to the provincial governments.

It is not easy to imagine, certainly not for me as a southerner, what life might be like in the far north under INF economics. But sooner or later, surely, northerners as well as southerners are going to have to bite the frugality bullet. Neither group would seem to have anything to gain in the long run by pretending otherwise or by procrastinating. Those who live in the north and who know the north intimately can best sculpt northern frugality. May their vision be apt and clear.

7

The Transition

"The transition to a time when oil is restricted to a much smaller range of uses than today is inevitable; but it will be difficult to manage."

Gerald Foley (1976, 1981)

"If the taxes on energy should go up overnight ... , they cause inflation, dislocation, and chaos. But if green taxes on energy are applied over a twenty-year period, producers and consumers have ample time to adapt, plan, and reinvent."

Paul Hawken (1993)

"We cannot precipitately abandon our primary energy sources without profoundly reshaping our way of life. But we also need to begin, with vigor and determination, the inevitable transition to the post-fossil fuel world."

Vaclav Smil (2008)

Once we have made the decision to switch over to Intelligent National Frugality (INF) economics, we will have to work out the details of the transition arrangements. That will be no small undertaking. The biggest challenge, however, will already be behind us. In the words of Bill McKibben, "The most difficult transition is a mental one: from a world where the answer to every problem is growth to a world where we have to be mature and accept the idea that we're going to have to focus on resilience, stability, and security."

To keep things in perspective, it is worth remembering that the changes in modern society during the past half-century or so have not exactly been minor. Consider what has happened to inter-urban passenger rail travel and to farming, for example. During the coming half-century, there is surely every likelihood of change on a similar scale regardless of whether or not we adopt INF economics. Our only choice for now would seem to be between change in a desirable direction and change in an undesirable direction. I cannot imagine how our present voracious appetite for energy and for other raw resources could ever co-exist with a healthy degree of social and economic and environmental stability. First we have to bring our energy consumption under control. Only then can we hope to achieve a more reasonable balance between change and stability.

Concerning transition arrangements, we will need to decide first of all how quickly or slowly the transition should take place. If it is too sudden, then wrenching disruption may occur in our individual lives. But if it is too gradual, then our present problems may continue to worsen unnecessarily. There is also a further consideration. If we delay too long before even beginning, then sufficient time for a comfortably gradual transition may no longer be available. Time does not appear to be our side in this matter.

Should the transition be spread out over five years? Ten years? Fifteen? Twenty? Any such number would presumably be satisfactory, provided that it received a broad measure of democratic support and provided that Gaia (see the Introduction) and available supplies of natural resources were willing to cooperate.

As regards basic transition numbers and durations, here are two examples:

1. Suppose that a nine-year transition is to begin on January 1st, 2021 and end on December 31st, 2029. In year one

(the year 2021) of this example, 90% of general government revenue would be obtained from the old taxation system while 10% would be obtained from the new tax on fossil-fuel energy. In year two (2022), 80% of revenue would come from the old system and 20% from the new, and so on. By year nine (2029), only 10% of revenue would come from the old system while 90% would come from the new. At the beginning of the tenth year, on January 1st, 2030, the final remnant of the old system would be abolished and INF economics would be fully established.

2. Suppose instead that a nineteen-year transition period is to begin on that same January 1st, 2021. It will then end on December 31st, 2039. In year one (2021) of this example, 95% of general government revenue would be obtained from the old taxation system and only 5% from the new. In year two (2022), the figures would change to 90% and 10% respectively. And so on. By year nineteen (2039), only 5% of revenue would come from the old system while 95% would come from the new. And on January 1st, 2040, the transition would be complete and INF economics would be fully established.

As the above two examples show, the main framework of the transition plan can be based on simple arithmetic. Almost certainly, however, this main framework will have to be reinforced by a number of supplementary measures. The purpose of these supplementary measures would be (a) to bring to an immediate halt, rather than to a gradual cessation, some of the least desirable economic activities that we are currently engaged in and (b) to institute immediate improvements in those areas (notably health, safety, poverty, unemployment and the environment) where delay is unconscionable.

Nuclear energy is a case in point. It might well make sense to close down some or all of our nuclear power stations immediately, rather than phasing them out over a period of several years. The idea would be to bring to a halt as quickly as possible the further production of long-lived radioactive wastes and to minimize the likelihood of any future nuclear accident.

Large-scale construction projects are another case in point. A few may continue to make sense under INF economics, but many will not. Most of the latter, one hopes, will be brought to a halt by market forces as soon as, or even before, our adoption of INF economics becomes official. But the forces of inertia can be quite strong. Our federal and provincial governments should therefore be prepared to legislate an immediate halt to any large-scale construction project likely to end up as a burdensome liability for future generations.

As for unemployment, two possibilities need to be considered. First, there may well be significant numbers of unemployed people even before the transition period begins. Secondly, in the early stages of the transition, old jobs may tend to disappear faster than new ones would normally be created. For both reasons, a contingency plan for dealing with transitional unemployment should be prepared.

Any such contingency plan should be designed to mesh well with INF economics. After all, if the ultimate goal is to move forward, then one does not normally begin by moving backward.

Here, then, is a brief restatement of what INF economics is all about. It is about stopping all generation of electricity from nuclear energy. It is about consuming, with each passing year, less and less fossil-fuel energy. It is about restructuring our whole economy so that energy in general can get used far more efficiently and far more frugally than at present.

More particularly, in as many different situations as possible, INF economics is about turning first to human energy and saying, "This job that we are now discussing is being offered to you first. If you want it, please take it. But if some other job holds more appeal for you, then please take that other job instead."

Next, INF economics turns to various other forms of renewable energy and says to them, "Having been turned down by human energy, this job is now being offered to you fellows. Are there any takers?"

Finally, if no forms of renewable energy step forward, INF economics turns reluctantly to fossil-fuel energy: "For the time being, we need some help badly enough that we are prepared to pay through the nose for it. So this job gets allocated to you. But because you cost so much, we will be trying very hard to disemploy you, either by giving the job to someone else or by eliminating it entirely. We will not be feeling any sentimental attachment to you at all. Sorry."

With that overall picture of INF economics in mind, let me offer a few suggestions as to what sorts of measures might get included in the contingency plan for preventing transitional unemployment.

Suggestions

My first suggestion concerns automobile repairs. Like other modern industrialized countries, Canada has an extremely broad and deep pool of competent automobile mechanics. One way to increase the number of jobs for such people throughout the whole country is (a) to remove all sales taxes and value-added taxes on used motor vehicles and used automotive parts (and possibly even on new automotive parts) and also on labour charges for automotive repair and maintenance work and (b) to compensate for the lost tax revenue by imposing a surtax on the full retail price of all brand new motor vehicles (imported as well as domestic). I realize that older vehicles often produce more air pollution than newer vehicles, and I am not suggesting that that problem should be totally ignored. But neither should we ignore the fact that the actual manufacture of a brand new motor vehicle also produces a significant amount of pollution and resource consumption.

If this first suggestion were implemented, the effect would be to discourage purchases of new vehicles and to encourage purchases of the human energy required to maintain existing vehicles in good running condition as long as possible. Since the manufacture of new vehicles is nearly always a highly automated process, the job gains should well outweigh the job losses. Moreover, many of our new vehicles are imports; in the case of reduced purchases of imports, the job losses would nearly all occur elsewhere, while the job gains would still occur here in Canada. As an added bonus, these job gains would be spread across the whole country rather than being concentrated in a few urban manufacturing centres.

In the case of passenger cars, one could go even further by imposing a very large surtax not on the total price of the new car but rather on the amount by which the total price exceeds the price of, say, a typical new basic compact car. (Special allowances could be made where the extra price is primarily attributable to fuel-efficiency features of the car in question.) Such a surtax would penalize only the buyers of luxury cars. And since most luxury cars in Canada are imported, probably an even greater percentage of the new work allocated to Canadian auto mechanics would represent work taken away from foreign energy (human, nuclear, hydroelectric and fossil-fuel). This may seem like a cynical shifting of our problems onto other countries, but I disagree. Charity begins at home, after all, and there is nothing to stop any other country from adopting INF economics whenever it wishes.

The above ideas do not have to be confined to the automotive industry. Any taxes that provide a direct incentive to be wasteful should be immediately abolished, including (a) all taxes on used merchandise of any kind, (b) all taxes on labour charges incurred for any kind of repair or maintenance work and (c) all taxes on labour charges incurred for any other purpose.

A second suggestion concerns employment related to various energy improvements. Many houses, for example, might be excellent candidates for extra insulation and for the installation of photovoltaic panels to convert incoming sunlight into electricity. A carefully planned government program that offered attractive loans and grants and useful advice might encourage large numbers of homeowners to go ahead and hire competent contractors and installers to improve the energy performance of their homes. This idea has been tried before, but not in the context of a planned transition to INF economics or something comparable.

A third suggestion concerns employment related to the railways. As our esteemed descendant noted in Chapter 3, there is every reason to expect a second golden age of railways in the years to come. Admittedly, INF economics will tend to discourage all mechanized forms of transportation, including railways. But the *relative* importance of rail transportation will almost certainly return to its level of former days. Moreover, Canada's population has grown considerably since the Second World War. The amount of railway-related employment should therefore be quite substantial in the new economy. Accordingly, there will no doubt be a number of railway-related areas in which labour-intensive projects could sensibly be undertaken soon after the transition period begins.

A fourth suggestion concerns job-training. The problem with job-training under conventional economics is that very often the jobs for which people are supposedly being trained are simply non-existent. Under INF economics, however, jobs and self-employment opportunities will be plentiful. Moreover, many of the kinds of skills that will be particularly useful in the new economy will not be difficult to predict. So even before some of the new jobs and opportunities (involving such activities as farming, cabinet-making and machinery-repairing) have come into existence, the government will be able to set up serious job-training programs geared to the requirements of the coming new economy.

A fifth suggestion is more general. When one thinks of all the repair and maintenance work that needs to be done in our society and when one bears in mind how frequently a stitch in time really does save nine, there is surely no good reason at all for allowing involuntary unemployment to persist during the transition period. Appropriate luxury taxes can be imposed, and extra government bonds can be sold, if necessary, for the purpose of keeping inflation in check and/or keeping government finances healthy.

Other points

Imports are another topic worth mentioning here. We may be tempted to buy too many imported goods during the transition period. If so, we would unwittingly be promoting unemployment in Canada since foreign jobs would be replacing Canadian jobs. In order to forestall this problem, it would be advisable to have in place, before the transition begins, suitable mechanisms enabling us to raise our import tariffs promptly and selectively as circumstances might warrant. (I have already discussed imports and tariffs in Chapter 5.)

All in all, successful management of the transition period will require a great deal of competence on the part of the government of the day. Those whose job it is to help design the transition economy will have to walk a fine line between (a) instituting so many extra rules, regulations, taxes and tariffs that the transition becomes far too disruptive and (b) standing aside and letting chaos occur.

Morale will also be crucial during the transition, since even with the best of intentions a certain amount of unfairness is probably inevitable. But if we genuinely believe in what we are doing and where we are going, and if our governments are on their toes to prevent the worst abuses and inequities and to provide a meaningful social safety net under all of us, then we can expect to emerge from the transition strong and undaunted and well prepared for the brand new world of Intelligent National Frugality.

8

Nuclear Power

"From a purely medical point of view, there really is no controversy: the commercial and military technologies we have developed to release the energy of the nucleus impose unacceptable risks to health and life. As a physician, I consider it my responsibility to preserve and further life. Thus, as a doctor, as well as a mother and a world citizen, I wish to practice the ultimate form of preventive medicine by ridding the earth of these technologies that propagate disease, suffering, and death."

Helen Caldicott (1978, 1994)

"Throughout my academic career I have declined to participate in research projects related to atomic energy, which I find an unforgiving and unforgivable technology ... I have always said that as soon as Canada decides ... to discontinue the building of nuclear reactors and to phase out existing ones, I would be happy to contribute all I have to addressing problems of nuclear-waste disposal."

Ursula M. Franklin (1990, 1999)

In this chapter, I offer my comments on three different questions relating to nuclear power, i.e. to the commercial use of nuclear energy for the purpose of generating electricity:

1. How affordable is nuclear power?
2. How dangerous is nuclear power?
3. How morally acceptable is nuclear power?

Nuclear affordability

Even under conventional economics, generating electricity in one of today's nuclear power plants is extremely costly. In the absence of large subsidies, the electricity so produced would essentially be unaffordable and hence unsaleable. Nevertheless, strong political forces have been promoting nuclear power. That is why, as Vaclav Smil has stated, "In many nations, nuclear energy generation has benefited from decades of generous public funding of basic research and operational subsidies."

Here, for example, is the first sentence of a *Globe and Mail* news story from September 24, 2003, written by Steven Chase: "Ottawa has doled out an extra $46 million this year to Atomic Energy of Canada Ltd. to help shore up development of new nuclear reactor technology—cash that's on top of the more than $100 million in annual federal subsidies the Crown corporation already receives."

Some nuclear subsidies are less obvious than others. One hidden subsidy takes the form of legislation that severely limits the legal liability of the operators of nuclear power plants. It works as follows. If and when a serious nuclear accident occurs, the resulting damage and destruction can be catastrophic, as we all know. Normally, the corporation or other entity that operates the reactor would be legally required to pay for all this damage and destruction. So in order to protect itself against huge damage claims, the operator would normally purchase liability insurance.

Insurance companies, however, are not stupid. Why would any insurance company run the risk of bankrupting itself by agreeing to provide accident liability insurance to the operator of a nuclear power plant? The insurance premiums charged would themselves have to be unaffordably large. More realistically, such insurance would simply not be offered at all.

But without access to liability insurance, why would any corporation risk bankrupting itself by operating a nuclear power plant? Who would want to buy shares in such a corporation? Or lend it money?

In order to resolve this perceived conundrum, countries that favour nuclear power have passed legislation severely limiting the legal liability of nuclear power-plant operators. As a result, ordinary citizens and homeowners are left legally unprotected in the event of a serious nuclear accident. Put another way, much of the net wealth of the general population is conditional upon no serious nuclear accident occurring. Put still another way, the general population is subsidizing nuclear power by carrying a financial burden that rightfully belongs on other shoulders.

Under Intelligent National Frugality (INF) economics, all subsidies (whether hidden or otherwise) to nuclear energy will be terminated, just as will all subsidies to fossil-fuel energy. Accordingly, legislation limiting the liability of nuclear power-plant operators will be repealed. That by itself will almost certainly result in the closing down of all our existing nuclear power plants and in the cancellation of any plans to build new ones.

In addition, INF economics will impose its new energy tax, expressed in dollars per unit of energy, on all nuclear energy resources (notably uranium ore and thorium ore) brought into the national economy. In other words, nuclear energy resources will be taxed on exactly the same basis as fossil-fuel energy resources. And the rate of the tax will be the same in both cases. But the consequences will be different. Nuclear power will almost certainly become immediately unaffordable and unsaleable, whereas fossil-fuel energy will continue to be used for quite some time, albeit with ever-increasing efficiency and frugality.

There is still more to the affordability question. Nuclear power involves a number of risks and dangers of various magnitudes, which I will be commenting on in the next section. Some, if not all, of those risks and dangers can be lessened if certain safety precautions are taken. But such precautions can be expensive. Just how expensive depends upon how much safety is desired. Every additional dollar spent on safety, however, has the effect (in the absence of additional subsidies) of raising the final price of the electricity generated. At some point, even with today's large nuclear subsidies, the increasingly expensive electricity would price itself right out of the market.

Under conventional economics, many countries, including Canada, have implicitly made the political decision that nuclear energy must be kept affordable. That is why nuclear subsidies have been provided. It is also why we have, in effect, given our nuclear designers and nuclear regulators implicit instructions to render nuclear power as safe as possible within the limits of affordability.

We couch this proviso in such words as "reasonable". "Take all reasonable precautions," we say to the nuclear experts, with everyone understanding that what we really mean is: "Don't make any decisions or recommendations that will cost too much!" Then, to make ourselves feel better, we use the word "acceptable" to describe those nuclear risks still being taken, as if the criteria for acceptability were a matter of technical expertise and therefore beyond debate.

Under INF economics, our general attitude towards nuclear power will almost certainly be quite different. We might even decide that nuclear power should be formally banned, regardless of the affordability question. But whether we formally ban it, or whether we allow the combination of INF economics and high safety standards to suffocate it, nuclear power seems virtually certain to become extinct.

Nuclear dangers

A number of different risks and dangers are associated with nuclear power. Some are minor and some are major. Some relate to terrorism. Some relate to the proliferation of nuclear weapons. And some relate to nuclear accidents of one kind or another. Already, the world has experienced three serious accidents at functioning nuclear power stations: Three Mile Island (Pennsylvania) in 1979, Chernobyl (Ukraine) in 1986, and Fukushima (Japan) in 2011.

When such accidents occur, sizable quantities of radioactive materials can escape into the environment (the atmosphere, the ocean, the groundwater, the soil). Radioactive materials give off a kind of radiation known as "ionizing radiation" which, depending upon its intensity and other factors, can cause serious illness and even death.

In order to think conceptually about the risks associated with nuclear power, I find it helpful to break down each risk into three components. The three components correspond to the answers to the following three questions:

1. What are the odds that the risk in question will actually materialize?
2. How much damage is likely to occur if the risk does materialize?
3. For how long will the risk continue to exist?

For each of the various nuclear risks being contemplated, the answers to these three questions will be different. For minor risks, the answers may not be troubling. But not all nuclear risks are minor.

We all know how to answer the three questions in a general way when we evaluate, say, the risk that the airplane on which we plan to fly will end up crashing.

But we are less familiar with the third of these risk components in situations where the risk goes on and on and on. Often, unfortunately, radioactivity does go on and on and on. The science underlying radioactivity is fairly well understood, and hence the so-called "half life" of any particular radioactive isotope is known. For some of the longer-lasting isotopes, the half life is measured in thousands of years. This means that certain risks associated with nuclear "spent fuel" waste will continue for thousands of years. None of the ordinary risks that we all deal with on a regular basis have anything like that kind of duration.

With risk duration in mind, we need to consider various kinds of serious nuclear accidents. In one kind, as for example with the Chernobyl accident, there will be a fairly sudden release of a very large quantity of radioactive materials from a functioning nuclear reactor. Radiation poisoning will then cause some people to sicken and maybe even to die. Some downwind crops may have to be destroyed. And some farms, villages, towns or even cities may have to be abandoned for weeks, months, years or decades. Basically, the risk of this kind of accident happening comes to an end when the nuclear reactor in question get permanently shut down. But the risk that the released radioactive materials may cause harm could continue for a very long time.

Another kind of serious nuclear accident, which may not yet have happened anywhere, involves the long-term storage of high-level nuclear waste. Some of that waste will remain dangerously radioactive for thousands of years. All during that long storage period, the danger exists that a significant amount of the waste will somehow escape from its sealed containers and find its way into, probably, the nearest groundwater. The consequences for anyone who comes in contact with that contaminated groundwater are not likely to be pleasant. Nor is avoiding such contact likely to be easy, especially since we have no way of knowing how much relevant technology or how much scientific knowledge about radioactivity will still be available in the distant future.

Because of risk duration, one cannot legitimately compare the dangers of nuclear power with the dangers of, say, commercial aviation. One cannot compare the risk of an airplane crash (where the risk lasts for a few hours in the case of a passenger or for several thousand hours in the case of a professional pilot) with the risk of being poisoned by radiation from escaped high-level nuclear waste (where the risk lasts for longer than all of recorded human history).

Still another nuclear risk needs to be noted, although here the science is uncertain and hence the degree of risk involved is itself uncertain. The question is this: How much harm, if any, is caused when human beings are exposed to low levels of ionizing radiation? Is there a threshold below which no risk of harm is involved at all? Or is all exposure dangerous, especially to children and pregnant women, and best avoided?

These questions are important because low levels of ionizing radiation are in fact given off at times in some of the processes involved in nuclear power production. But the questions are also highly controversial. If low levels of ionizing radiation are harmless, then we need not worry about them. But if they are not harmless, then precautions are in order. Such precautions could conceivably require that we turn our backs completely on nuclear power.

I myself do not claim to have any scientific knowledge on this matter. I do feel, however, that even with the best will in the world it is almost impossible to remain objective and unbiased in cases such as this. A great deal of money and a great many careers are at stake. It is not even clear where the burden of proof should lie. Should low-level ionizing radiation be presumed harmless until proven otherwise? Or should it be presumed harmful until proven otherwise? Not surprisingly, those who favour nuclear power tend to favour the harmlessness presumption, while we who oppose nuclear power tend to favour the opposite presumption. As the cynical witticism puts it, "Where one stands depends upon where one sits!"

Nothing is easier these days than to find support for the idea that low-level ionizing radiation is essentially harmless. On the other hand, I would like to offer a single sentence from the four hundred detailed pages of *No Immediate Danger? Prognosis for a Radioactive Earth* by Rosalie Bertell: "Most people are unaware of the fact that ionising radiation can cause spontaneous abortions, stillbirths, infant deaths, asthmas, severe allergies, depressed immune systems (with greater risk of bacterial and viral infections), leukaemia, solid tumours, birth defects, or mental and physical retardation in children."

Nuclear morality

Quite apart from questions of economics, nuclear power must be weighed against our society's professed standards of moral behaviour. If nuclear power fails to meet those standards, and in my view it does fail, then we should reject it no matter how affordable it might seem.

As I see it, the key moral problem (or at least one of them) here centres around our relationship to our descendants. What moral obligations are we under with respect to those who come after us? Is it acceptable if we kick them in the teeth, so to speak? Does the Golden Rule ("Do unto others as you would have them do unto you") apply towards the unborn? Is it fair if one generation or one group of generations gets all the benefits of some activity while knowingly leaving major costs and/or risks to be borne by future generations?

When these questions are posed in the abstract, with no specifics attached, we probably all agree as to how they should be answered. Once nuclear power enters the picture, however, we seem to lose the courage of our convictions. We find that we want to have our cake and eat it too. We still want to enjoy the perceived benefits and attractions of nuclear power, but at the same time we want to believe that future generations will in no way suffer on account of our voracious energy appetite. Accordingly, we want to believe that our nuclear experts will come up with some technical "solution" to the problem of what to do with our high-level nuclear waste. Once that "solution" has been found, we will thankfully implement it and then continue enjoying our energy binge. Or so we imagine.

We seem determined not to let ourselves acknowledge the obvious: No nuclear expert is ever going to be able to guarantee that such-and-such a method of high-level nuclear waste disposal will cause no harm to anyone during the long, long period of dangerous radioactivity. Moreover, no social scientist is ever going to have any way of knowing how civilized or uncivilized, how rich or how poor, how technologically sophisticated or unsophisticated, our descendants are likely to be a thousand years down the road.

Using softer language than mine, Edward S. Cassedy and Peter Z. Grossman comment on these points as follows: "Since the impacts may come millenia in the future, we are dealing with an ethical issue of vast intergenerational dimensions—one where the very societies that may be affected are unimaginable to us. This only compounds our dilemma, in deciding *to what lengths present-day society should go to limit the risks to its descendants thousands of years into the future."* (Emphasis in the original.)

Joanna Macy's language is different again: "If our long, ongoing evolutionary journey were real to us, if we felt the aliveness of our planet home and a living connection with those who come after, would we still want to sweep these [radioactive] wastes under the rug, hide them like a secret shame and go on about our business as before?"

In expressing opposition to nuclear activities, we males seem to have trouble matching the visceral disapproval that emanates from so many members of the fair sex!

I will end this section by summarizing my own point of view. We are under no compulsion to select economics as the basis for our acceptance or rejection of nuclear power. Moral considerations can and should trump economics. Under INF economics, however, there would almost certainly be no market for nuclear power even if moral problems were totally absent. Under INF economics, therefore, there will be no dilemma and no need for soul-searching. Virtue will cost us nothing! But unfortunately we will still have to deal somehow with all the nuclear waste already produced.

Nuclear power and greenhouse gases

In recent years, nuclear proponents, including newly convinced nuclear proponents, have been putting forward the argument that nuclear power is indispensable if we want to significantly reduce our total emissions of greenhouse gases. For my discussion of that point, please see Chapter 11.

Part 3: Energy and the Environment

9

Environmental Overview

"It is the sheer scale of our energy use that may trigger planetary responses not seen at lower levels of aggregate consumption. Those responses, directly or indirectly, will have various adverse effects on human population health."

A.J. McMichael (1993)

"The same fossil fuels that contribute to the greenhouse effect when we burn them don't stop there. They do us damage in other ways as well. They pollute the air we breathe. They destroy forests with acid rain. They dirty our homes, and our lungs, with soot. They corrode the stone in ancient monuments and kill the fish in ancient lakes. They shorten people's lives and damage their health. They sicken children, and in some parts of the world they kill them."

Isaac Asimov and Frederick Pohl (1991)

In this book, I take it for granted that environmental degradation constitutes a matter of serious concern to the reader. As René Dubos wrote as far back as 1968, "All thoughtful persons worry about the future of the children who will have to spend their lives under the absurd social and environmental conditions we are thoughtlessly creating."

The argument that we cannot afford to have a healthy environment is in my view as misplaced as the argument that we cannot afford to have a healthy population. Admittedly, we cannot afford to spend annually a million dollars on the health care of each and every member of society. And admittedly, we do

111

not have unlimited funds to spend on cleaning up the environment. But the key to both kinds of good health surely lies essentially in prevention. We therefore need to have a clear understanding of what sorts of activities actually bring about environmental degradation. Then we can take steps to try to eliminate those activities or at the very least substantially reduce their frequency and intensity.

The energy link

Two words, I think, sum up the prime cause of environmental degradation. Those two words are "energy consumption". Indeed, most instances of environmental degradation can be traced to unwise and excessive energy consumption on the part of dear old *Homo sapiens*.

(Strictly speaking, energy cannot be consumed; it can only be degraded until sooner or later it is no longer capable of performing anything useful. But in this book, I follow normal conversational usage in saying "energy consumption" instead of the more accurate but cumbersome phrase "consumption of energy usefulness".)

To be sure, volcanic eruptions and crashing meteorites and violent storms and so on can cause extensive temporary environmental damage. And no doubt most of the world's deserts owe their essential desertification to natural causes. But the kind of ongoing environmental degradation that threatens to cause us all so much anxiety and grief and misery can clearly be traced to the quantities and characteristics of modern energy consumption by human beings.

In the face of this link, however, conventional economics reacts very differently from Intelligent National Frugality (INF) economics. Conventional economics tends to assume that we human beings have enormous energy needs and that no economic system can be considered satisfactory unless it makes a serious effort to meet those supposed needs. Accordingly, conventional economics tends to turn a blind eye toward the link between energy consumption and environmental degradation, although a certain amount of hesitant eye-opening has admittedly begun in recent years.

INF economics, by contrast, will keep its eye firmly fixed on energy restraint at all times. As a result, the overall risk of serious environmental harm will be considerably lessened and the opportunity to deal with incipient environmental problems before they get out of hand will be considerably enhanced.

Many of the individual links between energy consumption and environmental degradation are known to us all, either because we can perceive them directly or because they have become common knowledge. Everyone knows, for example, that all nuclear power stations inevitably produce significant amounts of radioactive nuclear waste. Everyone also knows that the exhaust systems of gasoline and diesel engines emit not only such air pollutants as carbon monoxide (CO) and oxides of nitrogen (NO and NO_2) but also the greenhouse gas carbon dioxide (CO_2). And we have all seen ugly black smoke emerging from industrial smokestacks.

In the case of deforestation, gasoline-powered chainsaws clearly play a major role. Admittedly, deforestation can occur without chainsaws. Trees simply have no defence against axes and human-powered saws, let alone against chainsaws. Moreover, trees themselves constitute enormous stores of chemical energy and hence are quite capable of burning themselves up once a forest fire has been ignited. In our day, however, chainsaws and skidders and logging trucks and road-building bulldozers have all put fossil-fuel energy to work at the task of converting living trees into marketable raw materials. As a result, massive deforestation is occurring around the world.

In turn, deforestation often leads to further environmental degradation such as (a) soil erosion on hillsides and (b) loss of essential plant nutrients in tropical forests, not to mention (c) the net conversion of wood into atmospheric carbon dioxide (CO_2).

In similar fashion, the world's fishing fleets, powered by vast quantities of fossil-fuel energy, are slowly converting the bounty of the oceans into fading memories.

In agriculture as well, a strong connection exists between large-scale energy consumption and serious environmental degradation, with the latter notably involving excessive loss of topsoil. Wes Jackson has pointed out that so-called primitive peoples, whether in the distant past or more recently, relied mostly on human energy for their hunting and food-gathering activities. As a result, little environmental degradation of a serious and long-lasting nature occurred. "[W]e were altogether limited in our destructive ability," Jackson writes. He further sums up his point as follows: "No ecological ethic is necessary so long as the energy put into an area is spent through the arms and legs of the people of the area."

Jackson does not conclude, nor do I, that we moderns should therefore rely exclusively on human energy for all our food-producing and food-gathering activities. "But in the longer run," he comments, "a new agriculture will be necessary."

For those who have never heard of Wes Jackson, I will quote a sentence written by Carl N. McDaniel: "Jackson's work, as embodied in The Land Institute [cofounded by Wes and Dana Jackson in Kansas in 1976], is the harbinger of a twenty-first century agriculture that could not only save the remnants of good soil and wilderness, but also return much of our impoverished agricultural land to health by adopting ecological and evolutionary principles."

As for synthetic chemicals, they—and the pollution and contamination in which they are involved—have the distinction of being doubly linked to the consumption of major energy resources. First, the modern chemical industry consumes large quantities of energy in its manufacturing operations. And secondly, the industry needs a raw material or feedstock out of which to manufacture its various chemicals; the feedstock selected nowadays nearly always consists of a fossil-fuel energy resource such as oil or natural gas or a derivative thereof. So if under conventional economics the price of fossil-fuel energy resources is heavily subsidized (especially as seen from the point of view of INF economics), it follows that the price of synthetic chemicals is subsidized twice over. (I will have more to say about synthetic chemicals in Chapter 22.)

As all these examples show, behind nearly every instance of modern environmental degradation, at least in the industrialized countries of the world, lies a significant consumption of fossil-fuel and/or nuclear energy resources. Surely, therefore, the most important key to dealing with environmental degradation in the industrialized countries is to bring our energy consumption under control. That is indeed the approach taken by INF economics.

Nothing in INF economics, however, will prevent the government of the day from legislating whatever regulations or restrictions or prohibitions or legal liability it considers appropriate in order to provide additional protection to the environment. The same is theoretically true under conventional economics. But it is much easier to paddle downstream than upstream. As I argue throughout this book, it will be much easier to safeguard the environment under the environmentally friendly price structure of INF economics than under the environmentally hostile price structure of conventional economics.

Two problems, one remedy

Is it just a coincidence that INF economics will stand us in excellent stead with regard both to our environmental problems and to our unemployment and underemployment problems? Not at all. Both sets of problems share a common cause and therefore both share a common remedy. The common cause can be traced to our excessive and inappropriate consumption of energy. That excessive and inappropriate consumption of energy can in turn be traced to the inappropriateness of the relative prices that conventional economics imposes on the various kinds of commercially used energy in our society, including commercially used human energy.

To sum up, conventional economics steals energetically with both hands and both feet. With its left hand, it steals good jobs away from human beings. With its right hand, it steals good health away from the environment. And with its feet, it wastes enormous quantities of coal, oil and natural gas.

Overpopulation

At this point, many readers may be thinking, "How can anyone possibly discuss modern environmental degradation without making major reference to the problem of overpopulation? Surely our excessive consumption of energy can be attributed as much to our excessive population size as to our excessive per capita energy consumption!" My response follows:

1. This is a book about economic policy choices and about energy restraint. Regardless of how we choose to react to

116

the overpopulation problem, we still have to decide what kind of national economy we wish to create for ourselves. I agree that, other things being equal, a large population consumes more energy and inflicts more environmental damage than does a small one. But that only reinforces the arguments in favour of intelligent frugality.

2. A sensible economic policy in no way precludes a deliberate population policy. INF economics, for example, in no way precludes a policy whereby a substantial life-annuity, beginning on the annuitant's fiftieth birthday, would be paid by the government to all women who, at the time of that fiftieth birthday, have not given birth to more than two children.

3. Curiously, whenever a country starts producing fewer children, a great many commentators express concern over how much difficulty the younger generation is going to have in supporting all the retirees in the older generation. Such commentators therefore tend to advocate policies promoting more births or more immigration or both. But surely a natural drop in the birth rate constitutes by far the most humane scenario for a reduction of population size. I see no ethical justification for taking deliberate steps to leave some future generation with an even more serious overpopulation problem than our own. I agree completely with the following comment by Paul and Anne Ehrlich: "That halting population growth inevitably leads to changes in age structure is no excuse for bemoaning drops in fertility rates, as is common in European government circles. Reduction of population size in those over-consuming nations is a very positive trend, and sensible planning can deal with the problems of population aging."

Blind spots

We all tend to have our own blind spots when it comes to the connection between energy consumption and environmental degradation.

Suppose, for example, that I live in the middle of a big city and that on winter weekends I enjoy getting away from the city and doing some serious cross-country skiing on remote wilderness trails. Suppose further that I do not have much of a warm spot in my heart for snowmobiles. Their noise and fumes and constant zooming around offend me whenever our paths cross. So I tend to be highly critical of both the energy waste and the environmental degradation that I attribute to snowmobiling. But what I fail to see, because of my all-too-human blind spot, is that on those very weekends when I am out skiing, I may in fact be consuming more gasoline and producing more air pollution and more carbon dioxide (CO_2) emissions than any of the snowmobilers I encounter. Having driven my weighty minivan or my weighty sport utility vehicle a good many kilometres to and from the ski trails, I may indeed be the environmental booby of the day.

A similar blind spot could apply to a canoeist surrounded by motorboats, to a hiker surrounded by off-road motorcycles, to a hang-glider surrounded by conventional airplanes or to a walking golfer surrounded by electric golf carts.

Sometimes, especially when the energy consumption involved is indirect, our blind spots can attain surprising size. Consider, for example, the modern version of professional team sports, which generally involves a great deal of air travel on the part of the athletes, coaches and managers. The associated energy consumption and environmental degradation must be quite significant. And yet, watching the performance on our living-room TV or sitting in the ultra-modern arena, we spectators tend to feel quite innocent of any environmental transgressions.

INF economics will improve everyone's vision enormously in these matters. It will do so by the simple expedient of ensuring that we each pay a substantial amount of money for every unit of fossil-fuel energy that we personally are responsible for consuming, regardless of whether that consumption is direct or indirect. One gets the proper focus quite quickly when a substantial outflow of cash is taking place.

In the case of direct purchases of gasoline, for example, every purchaser will be immediately aware that he or she has just paid a large tax (a) for the privilege of contributing further to the depletion of our stock of exploitable fossil-fuel energy reserves and (b) for the privilege of imposing additional stress on various components of the environment. The same will apply—to the extent that fossil-fuel energy consumption is involved—to the purchase of bus tickets and train tickets and airplane tickets and to the payment of taxicab fares.

As for all the fossil-fuel energy used to ferry professional sports teams back and forth across the country or across the continent, we do not need to know right now which group of people will be paying the bill. It might be the spectators in the arena. It might be the television viewers. It might be the customers of some commercial sponsor. It might be some combination of all three. But if all three decline to pay, then professional team sports as we know them will disappear. Why should any form of professional entertainment not disappear if, financially speaking, it cannot stand on its own two feet?

Put more generally, why should any environmentally unfriendly activity of any kind be subsidized? INF economics will insist that we all pay a realistic price for fossil-fuel energy. (For more on this point, see Chapter 16.) As a result, it is unlikely that any of us will fail to see and understand our own role in the overall energy and environment picture.

Governments

One might wonder whether INF economics, in removing blind spots from the eyes of the citizenry, will not simply shift the blindness to the government itself. After all, any money paid out in taxes by taxpayers will get received as incoming revenue by the government. What incentive will induce the government itself to consume fossil-fuel energy wisely, given the likelihood that both expenditure and revenue figures might rise more or less equally if energy restraint were lacking?

Fortunately, the desired incentive will still be quite strong. Suppose, for example, that under INF economics the federal department of citizenship and immigration decides to indulge in some fossil-fuel energy extravagance by frequently flying a great many of its employees here, there and everywhere. Such extravagance will of course cost that department quite a lot of money because of the substantial tax on fossil-fuel energy. But where will all the proceeds of this energy tax go? The money will go, first of all, to the government's consolidated revenue fund, since that is where nearly all government revenue initially goes. From there, the money will be divided up among various government departments. And in a federal system of government such as Canada's, some of the money will be allocated to the various provincial and territorial governments as well. Not more than a tiny fraction of the original amount paid out by the department of citizenship and immigration will ever find its way back to that same department. And even then, that tiny fraction will have to be shared between the immigration branch (which, let us say, made the original decision to abandon frugality) and the citizenship branch (which, let us say, did not).

Under INF economics, then, any civil servant or government minister who makes a decision to purchase extravagant amounts of fossil-fuel energy on behalf of his or her government department will in effect be donating a substantial amount of departmental money to the consolidated revenue fund and will probably never regain control of the money so donated. That is not the sort of donation that civil servants and government ministers are considered to enjoy making. Hence the incentive to be frugal and to use energy efficiently will be almost as strong for the government as it will be for businesses and consumers.

An immense alien creature

To end this overview chapter on the environment, I would like to offer the following thought from David Suzuki: "Suppose an immense alien creature came to Earth in a spacecraft from a distant galaxy and began to tear up the planet. Striding at the rate of a step a second, with each footstep crushing an acre of forest, belching poisonous gases into the atmosphere and excreting toxins onto the earth and into the water, the monster would fill us with fear and galvanize all of humanity into an all-out effort to vanquish it. The impact of that hypothetical alien mirrors the actual destruction being wreaked on Earth by us. Seen this way, it becomes obvious we need to be united as a species to stop the deadly assault."

10

Selected Energy-Environment Links

As I argued in the previous chapter, unwise and excessive energy consumption constitutes the key to most kinds of human-caused environmental degradation. I will continue to develop that argument in the present chapter.

The links between energy consumption and environmental degradation are by no means always direct. Nor is each individual link always independent of all other links. On the contrary, a great deal of indirectness and a whole web of interconnectedness are involved here. That is one of the reasons why Intelligent National Frugality (INF) economics focuses on raw energy resources in the first place. Such a focus enables the taxation system to get in on the ground floor, so to speak, before all the indirectness and interconnectedness render the overall picture impenetrably complex.

Precisely because it gets in on the ground floor, INF economics will be able to use to its own advantage all the indirectness and interconnectedness that I have just mentioned. A pro-environment bias will thus work its way into every nook and cranny of the whole economy.

In order to provide a number of specific examples of this indirectness and interconnectedness, the present chapter will examine four representative energy-environment topics, beginning with garbage and waste.

Garbage and waste

"In a closed system there is no such thing as disposal."

Ralph O. Brinkhurst and Donald A. Chant (1971)

Neither human beings nor any other creatures are biologically capable of remaining alive—or of dying, for that matter—without inflicting a certain amount of waste upon the environment. Normally, however, nature recycles such waste. Any environmental damage involved tends to be both minor and temporary. In the case of a colony of great blue herons nesting together in a collective heronry, for example, the gradual accumulation of droppings ends up killing more and more of the on-site vegetation until eventually the colony moves elsewhere and the site slowly recovers.

Unfortunately, the waste-related environmental damage caused by modern human behaviour cannot be described as minor and temporary. Under the conditions of modern industrial civilization and conventional economics, we have acquired the habit of producing enormous quantities of garbage and waste, much of which cannot easily be recycled at all. Consequently, (a) we are running out of suitable "sinks" in which to deposit our waste harmlessly and (b) our waste is causing various kinds of serious environmental degradation such as air pollution, fresh-water pollution, ocean pollution, soil pollution, habitat destruction, stratospheric ozone depletion, increased atmospheric concentrations of greenhouse gases and so on.

In many of these waste-related environmental problems, the energy link is readily apparent. All the environmental problems associated with nuclear waste, for example, are clearly linked with nuclear energy. Similarly, the ocean pollution caused by the leaking of crude oil and other petroleum cargoes from damaged ships such as the Exxon Valdez has an obvious link with fossil-fuel energy. So do the air pollution and the emissions of carbon dioxide (CO_2) that occur when fossil fuels get burned; this includes the situation where the petroleum industry deliberately burns off, as waste products, any unwanted or unusable hydrocarbon gases. As for the tailings ponds associated with the exploitation of Alberta's oil sands, Andrew Nikiforuk describes them as "one of the world's most fantastic concentrations of toxic waste."

What about solid waste of the kind that ends up in municipal incinerators or in so-called landfill sites? Consider, in particular, all the waste materials—wood, paper, cardboard and so on—that originate in trees. Such materials have both a direct link and an indirect link with energy.

The direct link lies in the fact that tree-derived waste generally contains a portion of the tree's original solar-derived chemical energy. That is why such waste is capable of burning. It is also why, when buried in a landfill site and thus deprived of contact with oxygen gas (O_2), such waste is capable of undergoing anaerobic decomposition and giving off methane gas (CH_4). When significant quantities of such methane escape to the atmosphere without being intercepted, they constitute an environmental hazard as regards both air pollution and atmospheric greenhouse-gas buildup.

As for the indirect link, it lies in the fact that trees do not turn into commercial waste products without help. Considerable amounts of energy from external sources are needed. The trees have to be felled, skidded, loaded, trucked, unloaded and processed, with energy being required at every step. Admittedly, sawdust and scrap wood can provide some of this needed energy. Nevertheless, without external sources of energy, trees stay in the forest. They might burn up right in the forest, by converting their own stored chemical energy into heat. But they will not transform themselves into wooden furniture or construction lumber or wooden toys or junk mail or cardboard packaging or yesterday's newspaper. Not without external energy.

Now if, under conventional economics, we find that our production of tree-derived waste is absurdly excessive, does it not make sense to wonder whether most of the external energy consumed in the corresponding production processes might be severely underpriced? That would certainly explain both why we consume so many trees in the first place and why we make so little effort to prevent tree-derived products from becoming tree-derived waste.

After all, demand tends to go up when price comes down. Demand, in this case, means demand by the forest-products industry for external energy for use in converting living trees into commercial goods. Moreover, concern for such qualities as durability, re-usability, repairability and recyclability tends to evaporate when costs are low and supplies seem plentiful. In short, an abundance of cheap energy leads almost inevitably to mountains of tree-derived trash.

As with tree-derived trash, so with petrochemical-derived trash (notably plastic trash) and so also with mineral-derived trash (notably metal junk). In all these cases, production of the material that ends up as trash—whether it be leftover trash from manufacturing processes or the remains of what was once a marketable commercial product—does not cost very much money under conventional economics. The reason it costs so little is that conventional economics assigns very low prices to most of the external energy involved. Such external energy is used to convert (a) a portion of the original forest or (b) a portion of the underground petroleum deposit (which is itself available at absurdly low prices) or (c) a portion of the underground deposits of mineral ore, as the case may be, into whatever it is that ends up as unwanted trash.

Human energy too deserves mention. Its very *high* price under conventional economics aggravates the trash problem still further, both directly and indirectly.

In the direct situation, overpriced human energy is often too expensive to be used for the purpose of reconverting trash into something useful. Sorting, trimming, removing old nails, cleaning, repairing and so on are often perfectly feasible, but only if done with a large input of human energy. No wonder we find ourselves sending our unwanted old ships, for recycling, to poorer countries where human energy is cheap and where safety and environmental standards leave much to be desired.

In the indirect situation, people who design such items as radios and automotive parts and hundreds of other products are very often discouraged from incorporating ease of repair into their designs. They are discouraged by the knowledge that under conventional economics human energy is generally far too expensive to be allocated to the kind of repair work in question. Why go to the trouble and expense of incorporating a design feature that will probably never get used or even appreciated?

There is also what I would call a "downstream" link between our trash and our energy consumption. Because our trash volumes under conventional economics are so large, we employ a sizable fleet of fossil-fuel-powered trucks to transport our trash to its assigned destination: a landfill site, an incinerator or a recycling facility. Moreover, our highly automated recycling plants make liberal use of electric motors (to power conveyor belts) and other energy-intensive machinery. So in many cases our trash-recycling efforts would seem to compound our energy-consumption excesses.

Some readers might argue that even more energy consumption would be necessary if no recycling took place. Under conventional economics, that might well be true. But under INF economics, far less energy will be required in total because far less trash will get produced in the first place and because all recycling efforts will themselves be carried out in a decentralized and energy-efficient manner.

Finally, since our mountains of tree-derived trash have links not only with external sources of energy but also with other tree-related environmental problems, here is a 1998 comment from Elizabeth May: "[A]cross Canada, too little attention is paid to the direct polluting impact of the forest industry. In assessing the real cost of society's addiction to wasteful paper consumption, every part of the process has to be factored in—from clear-cutting and herbicide spraying of plantations to pulp mill effluent and oriented strand board off-gassing in homes."

Concrete and steel

"Immense skyscrapers, bridges, and dams, drained swamps and deep-sea oil rigs, are a testimony to our engineering skills. But too often we fail to consider a project's impact beyond its immediate locale or payoff, and we end up paying heavy ecological costs."

David Suzuki (1994)

In this second section, I would like to look briefly at some of the energy consumption and environmental degradation associated with the production and use of concrete and steel. Both materials are familiar to everyone. Both can be found not far from almost every human being living in the modern industrialized world. And both come at an environmental price that tends to increase sharply as the quantity of the concrete or steel involved goes up. For convenience, let us divide this environmental price into its "upstream" portion and its "downstream" portion.

Looking "upstream" from the point of view of a newly produced batch of steel or Portland cement (the latter is the energy-intensive ingredient in concrete), we see two things. We see large-scale consumption of non-renewable energy resources, and we also see significant environmental degradation. We see open-pit iron-ore mines, limestone quarries, blast furnaces, coal yards, tall industrial chimneys and so on. We see the unwashed face of raw industrial power. And we can imagine the billowing "clouds" of invisible carbon dioxide (CO_2) that emerge from those tall chimneys and increase the atmosphere's total concentrations of greenhouse gases. In fact, the cement industry emits significant amounts of carbon dioxide in two different ways: (a) by burning fossil fuels in order to create needed heat and (b) as a by-product of the chemistry involved in the production of Portland cement. (Efforts are currently being made to alter the chemistry involved so as to reduce the production of cement-related CO_2.)

Looking "downstream" from the production of concrete and steel, we once again tend to see the same two things: (a) large-scale consumption of non-renewable energy resources and (b) significant environmental degradation.

For a first example of this "downstream" phenomenon, consider nuclear power stations. Huge quantities of concrete and steel and energy are required for the construction of such stations. But the latter are of no value unless they actually operate, with consequent production of environmental hazards in the form of high-level and low-level radioactive wastes, not to mention the radioactivity dangers that might result from a nuclear accident.

For a second concrete-and-steel "downstream" example, consider the enormous concrete dams that we build from time to time in order to supply an abundance of kinetic energy to the turbines of very large hydroelectric generating stations. Here too, we see significant consumption of non-renewable energy, even though we like to think of hydroelectricity as being a renewable form of energy. As I mentioned in Chapter 2, large dams and large generating stations require large energy inputs in order to get constructed. Think of the energy required for all the blasting and earth-moving. Think of the energy required for producing all those cubic yards of concrete and for transporting them to their final destination. Nearly always, most of this needed energy can be traced to non-renewable sources. So a large hydroelectric generating station generally has to produce electricity for quite some time before it can be said to have repaid the energy "debt" incurred during the construction process.

As for environmental degradation, it too would be mostly associated with the initial construction, rather than with the day-to-day operation, of the dam and the generating station. First and foremost would come habitat destruction, owing to the massive flooding that normally takes place. Secondly, a great deal of air pollution and greenhouse-gas emission accompanies the construction activities. And now there are indications that the flooding process itself may cause mercury (Hg) pollution, apparently as the result, in the words of B. D. Roebuck, of "the action of bacteria on soils and plant material that contain mercury, specifically decaying trees, bushes, and plant matter in flooded reservoirs."

Two additional dam problems may arise as time passes, and each of them may have environmental ramifications. The first involves the gradual silting up of the new lake or reservoir located on the upstream side of the dam. If silting up does occur, not only does it detract from the hydroelectric performance of the generating station, but also it may deprive downstream regions (such as the Nile River delta in Egypt) of needed silt. The second problem is summed up in the following comment by Vaclav Smil:

"We are not designing large dams to last indefinitely—but neither are we building them in ways that could make it easier to decommission them in the future." Interestingly, however, the United States has already begun demolishing certain of its dams having negative environmental features.

Nor can we ignore the ugly line of transmission towers that often has to snake its way across the landscape in order to connect remote hydroelectric generating stations to urban population centres. Such transmission lines cause further habitat destruction as well as visual pollution. Moreover, we do not yet know what harmful human health effects, if any, might result from the high transmission voltages involved.

A third "downstream" example of energy-environment linkages involving concrete and steel can be seen in our modern superhighway systems with all their bridges and interchanges and their many parallel lanes of roadway. Here, both the "downstream" energy consumption and the "downstream" environmental degradation are easily pictured. The former centres around an extravagant-beyond-words consumption of motor fuel, while the latter comprises (a) habitat destruction, (b) air pollution, (c) atmospheric emissions of carbon dioxide (CO_2), (d) atmospheric emissions of harmful refrigerants from leaking automotive air-conditioning systems, (e) visual pollution and (f) noise pollution.

Many other examples of "upstream" and "downstream" energy flows and of consequent environmental degradation could easily be given, not just for concrete and steel but also for copper (Cu), aluminum (Al), tin (Sn), lead (Pb), petrochemicals and so on. In each case, the point would be the same: Under conventional economics, there is a marked tendency for energy consumption to feed on itself, more or less out of control. As a result, environmental degradation too would seem to be more or less out of control. The problem, in my view, lies squarely with the choices and assumptions inherent in conventional economics.

Let me return to concrete for a moment. Few of us purchase very much of it directly, certainly not on an annual basis. Nevertheless, the annual per capita production and sale of concrete in the modern industrialized world is very large. Somebody must be buying the stuff! The obvious explanation is that corporations and institutions and governments purchase large quantities of many materials, including concrete. What makes concrete rather unique as a material is that most concrete ends up in structures that are far too large to be purchased by private individuals.

Because most concrete is purchased by large entities, and because concrete is intimately linked with environmental degradation in a variety of different ways, it would appear that we private individuals cannot do very much to lessen concrete-related environmental degradation. In actual fact, however, we all wear two different hats, depending upon the circumstances. When we are out shopping, we wear our consumer's hat and we take the economy's price structure the way we find it. But when we are discussing what kind of an economy we would like to choose for our country, and when we are exercising our right to vote at the ballot box, we wear a different hat entirely. We wear our citizen's hat.

As citizens, we choose our government, which in turn chooses our economic system. The latter then establishes the basic pattern of our overall price structure, thereby determining to a large degree how wastefully or how frugally our society deals with all its resources, including both concrete and steel. So we need to distinguish clearly between (a) those situations where, as a consumer, one votes with one's purse or wallet and (b) those situations where, as a citizen, one votes at the ballot box.

In the case of items such as electric can-openers, for example, those of us who are not physically handicapped and therefore have no need of such items can vote with our purse or wallet by abstaining from buying them. But in the case of concrete and its associated environmental degradation, any votes cast other than at the ballot box will generally have, at best, only a minor effect.

Antibiotics and animal husbandry

"Despite these and a host of other examples of the transmission of antibiotic-resistant bacteria from meat, dairy, and poultry products to human consumers, the U.S. Food and Drug Administration, its counterparts in Europe, and the European Community (under the Maastricht Treaty) all failed to take actions that might have limited the use of antibiotics on animals."

Laurie Garrett (1994)

For a different kind of environmental degradation, but a kind that is no less real and no less serious than the kinds previously mentioned, consider the environmental impact of antibiotic drugs. Today's news media are telling us with increasing frequency and increasing alarm that more and more disease-causing bacteria are developing new strains resistant to antibiotics. The disturbing speed with which this development is taking place can apparently be attributed in part (note the words "in part") to the widespread use of antibiotics in modern farming.

The problem is that antibiotics do two different things simultaneously: (a) they kill bacteria and (b) they encourage, so to speak, unkilled bacteria to develop resistance to the antibiotic in question and sometimes to other antibiotics as well. Clearly, the more a given antibiotic successfully carries out the second of these two activities, the less able it will be to carry out the first.

133

Hence the title and subtitle of Stuart Levy's book *The Antibiotic Paradox: How Miracle Drugs Are Destroying the Miracle*. We human beings obviously have every reason to try to prevent, or at least slow down as much as possible, the second activity.

Everyone agrees that the only known way to inhibit the development of antibiotic resistance in bacteria is to limit the bacteria's exposure to antibiotics in the first place. That is why we are constantly being advised to avoid putting antibiotics into our body unless our medical condition so warrants. It is also why our farming practices ought to minimize the consumption of antibiotics by farm animals.

Under conventional economics, however, our use of antibiotics in connection with our farm animals is quite extensive, for two different reasons.

One reason is that we often raise large numbers of farm animals in crowded indoor conditions where bacterial illnesses can easily develop. Gene Logsdon makes the following observation: "My dairy cows, pasture-born, pasture-raised, and mother-nursed, have never gotten [the disease called] scours. Confinement cattle, on the other hand, must be pumped full of antibiotics to keep them healthy, resulting in the danger of bacteria becoming immune to antibiotics."

The other reason why farmers supply antibiotics to their animals is to promote rapid growth and rapid weight gain. In the words of Stuart Levy, writing about the American situation, "[A]ntibiotics are given to animals in small, subtherapeutic levels as a means to improve their growth." The daily amounts per animal are much smaller than in the case of illness treatment, Levy explains, but these "smaller doses are administered for longer periods of time, for weeks to months." A small daily dose multiplied by a large number of days leads to a substantial dose per animal. And a substantial dose per animal multiplied by a huge number of animals leads to an enormous overall total amount.

Michael Specter comments on this situation as follows: "Roughly three-quarters of the antibiotics consumed in the United States are fed to poultry, cows, and pigs, not to treat illness but as dietary supplements to promote faster growth. That saves the meat industry a lot of money; the sooner the animals reach a market weight, the sooner they can be slaughtered and sold. Until recently, the biochemical reasons for that weight gain, and its unsettling implications for humans, were murky. The new data suggest that even minimal exposure to antibiotics alters the gut bacteria of these animals, which may influence their ability to metabolize nutrients properly. As a result, researchers have concluded, both their body-fat percentage and their weight increase significantly."

Intelligent National Frugality (INF) economics, as our esteemed descendant suggested in Chapter 3, will revolutionize the vocation of farming. In a hundred different ways, it will tend to de-industrialize agriculture and to overturn modern agribusiness. Most farms will be far smaller than today, far more self-sufficient, far more attuned to natural processes, far less capital-intensive and far less energy-intensive. Dairy cattle, for example, will be raised in accordance with Gene Logsdon's principles, as summarized above. And chickens will spend much of their time outdoors with plenty of opportunity for finding bugs and other wild food and for breathing fresh air.

Under these new (but at the same time very old!) conditions, the health of most farm animals will be excellent and their illness-related need for antibiotics extremely small.

As for feeding subtherapeutic amounts of antibiotics to farm animals in order to speed up their growth, I foresee a complete cessation of this practice under INF economics. The following factors are worth noting:

1. Antibiotic drugs may become rather expensive, depending upon how much energy is consumed, directly and indirectly, in their development and manufacture.

2. Financial pressure to maximize rates of animal weight gain will be lowered because farms will typically possess a much greater degree of self-sufficiency than at present, will not be capital-intensive and hence will not be saddled with enormous bank loans.

3. With small-scale, decentralized meat production and butchering, meat customers will be more able (a) to ascertain details of the conditions under which the meat on offer has been produced and (b) to choose which farm or which kind of farm they wish their purchases of meat to be supplied from. Few customers will want even small traces of antibiotics in the meat they buy.

4. Perhaps most important of all, the absence of any ideological insistence on perpetual economic growth will encourage the whole of society to focus first and foremost on healthful nutrition and a healthful environment for both humans and farm animals. Governments will thus be likely to impose strict regulations concerning the use of antibiotics on farm animals. (Please re-read the Laurie Garrett quote at the beginning of this section.)

We may not be accustomed to viewing antibiotic-resistant bacteria as an example of environmental degradation. But if and when a good number of antibiotics do lose their effectiveness, our environment will become significantly more hostile to those of us unlucky enough to contract a serious bacterial illness.

Bulldozing the landscape

"Then the nightmare began. Trucks and bulldozers rolled in. Forests were felled, standing crops destroyed. Everything turned into a whirl of engineers and jeeps and cement and steel. Mohan Bhai Tadvi watched eight acres of his land with standing crops of sorghum, lentils, and cotton being leveled. Overnight he became a labourer."

Arundhati Roy (1999)

For many people, including myself, the damage and destruction that we have been inflicting on our landscape in recent decades constitutes one of the more saddening examples of modern environmental degradation. Such damage seems so unnecessary, so short-sighted, so wasteful, so destructive, so ugly and so wrong.

In order to witness what has happened, it is only necessary to visit the outskirts of almost any city or good-sized town in the whole of North America and no doubt elsewhere as well. There we encounter the typical commercial "strip" consisting of oversized advertising signs, mediocre architecture, acres of asphalt, motor vehicles everywhere, stewardship and domesticity almost nowhere. For some reason, we call such scenes "development". That kind of "development", I suggest, has the same economic cause as most other forms of environmental degradation: the malignant underpricing of nuclear and fossil-fuel energy. How else can one account for the commercial success of so much destruction of the natural landscape, so much consumption of natural resources and so little production of anything that human beings really need?

Consider the lowly bulldozer. No one will deny that bulldozers and other kinds of earth-moving equipment have made it laughably easy for us to convert rolling farmland into a huge shopping mall. Or wild wetland into a metropolitan airport. Or a healthy living woods into a giant amusement park. But why does the modern world manufacture so many bulldozers? And why do we use them so indiscriminately? The answer, as I see it, lies in the fact that under conventional economics one hour of a bulldozer's operating time has a very low cost when compared with the work performed. Who can resist such a bargain?

Not that bulldozer-work normally gets marketed to retail consumers. Rather, it normally gets marketed to businesses. The latter naturally expect to make a profit from their investment in bulldozer-work. In principle, I see nothing inherently wrong with that. But there can be underpriced bargains in investment expenditures, just as there can be underpriced bargains in consumer expenditures. Indeed, a particular unsubsidized investment expenditure priced at $1,000 might be totally unattractive from a business standpoint, whereas the same investment might become almost a guaranteed moneymaker if its price were to fall to $500 or $200 or $100. Governments often offer such carrots, as everyone knows, in attempts to attract new business investment. Not surprisingly, then, the demand for bulldozer-work picks up as the price goes down.

Under conventional economics and seen from the point of view of INF economics, bulldozer-work is tremendously underpriced. It is underpriced in two different ways in particular: (a) by virtue of the fact that the cost of manufacturing bulldozers is heavily subsidized and (b) by virtue of the fact that their operating costs are also heavily subsidized. Both sets of subsidies trace their origin to our heavy subsidization of nuclear and fossil-fuel energy. Moreover, the two bulldozer-work subsidies are so large that they do not even come close to being cancelled out by the overpricing of the tiny amount of human energy supplied by the bulldozer operator.

138

Once the cost of an hour's work by a bulldozer has been made dirt-cheap, instances of heartless destruction of landscape are bound to occur. No wonder Wendell Berry has expressed worry about the land just beyond the window of the room in which he has worked for thirty-seven years. "I have known this place all my life," he explains. "I long to protect it and the creatures who belong to it ... This is a small, fragile place, a slender strip of woodland between the river and the road. I know that in two hours a bulldozer could make it unrecognizable to me, and perfectly recognizable to every 'developer'."

Being a writer of fiction as well as non-fiction, Berry made essentially the same point by having the narrator of one of his short stories express her dismay in the following words: "Mr. Gotrocks hadn't any sooner paid his investment into it [i.e. paid for the farm that he was buying] than he hired a man with a bulldozer to smash the house and other buildings all to flinders, and push them into a pile, and set them afire. He pushed out every fence, every landmark that stood above the ground, every tree. A place where generations of people lived their *lives*. If they came back now, looking for it, they wouldn't know where they were."

James Lovelock generalizes that same sentiment: "We need a warning placed on every bulldozer, chainsaw, and on all large energy-using devices: 'Do nothing that would harm the Earth.'"

Under conventional economics, even further incentive to bulldoze arises from the fact that subsidized fossil-fuel energy is widely available to all of us for our cars and our various recreational activities. Large amusement parks, for example, have been bulldozed into existence because they benefit commercially from all the subsidized energy consumed (a) in their construction, (b) in their daily operation and (c) by the passenger cars and other motor vehicles that transport customers to and from the park.

As noted earlier, energy consumption under conventional economics tends to feed on itself, more or less out of control. With each extra mouthful, the resulting indigestion—from which we all end up suffering—becomes a little more serious, a little more painful and a little more dangerous.

Intelligent National Frugality (INF) economics

In the modern industrial world at present, we have a situation in which large-scale energy consumption is linked to serious environmental degradation through a vast web of interconnectedness. Now along comes INF economics. How will things change?

Conceptually speaking, INF economics can be thought of as comprising thousands and thousands of specific taxes on pollution, on habitat destruction, on greenhouse-gas emissions and so on. Indeed, almost every environmentally unfriendly activity imaginable will get a tax levied on it. In each case, the tax will get fine-tuned to the point where it is exactly proportional to the sum total of all the direct and indirect inputs of fossil-fuel energy that underlie the activity in question. And this magnificent state of affairs (if I may call it that) will require hardly any bureaucracy or paperwork at all!

Two points remain to be emphasized.

The first point is that these thousands and thousands of (conceptual) new environmental-protection taxes will not be small. In fact, when all added together, they will make up almost the totality of federal and provincial and territorial government revenues. As a result, virtually all of us—consumers, producers, governments, institutions—will change our economic behaviour substantially. We will not simply grumble a bit and then continue on as before. The new financial incentives and disincentives will be too large to ignore and too compelling to resist. And in virtually every single case, the consequent changes in our economic behaviour will lessen environmental harm and enhance environmental protection.

Put another way, INF economics will employ the technique of "downsizing" as its primary strategy for promoting environmental restoration and protection. But instead of downsizing by laying off human energy, as is sadly all the rage at present, it will downsize (a) by laying off all our nuclear power plants and related facilities (except those facilities dealing specifically with already-produced nuclear waste) and (b) by laying off a good portion of the gluttonous fossil-fuel energy consumption that is currently bloating poor Gaia's long-suffering abdomen.

My second point simply repeats what I have said previously. INF economics does not have to do the whole environmental-protection job all by itself. To whatever extent necessary, it can receive assistance from legislated regulations, restrictions, prohibitions and so on. Conceptually speaking, we first need to get the economics right. Then we can focus on tidying up any loose environmental-protection ends remaining.

11

Global Warming and Global Climate Change

"Of all the changes with which modern civilization threatens biospheric cycles, one is of by far the greatest concern, and it has become the most prominent environmental preoccupation of the waning twentieth century. This is the possibility that rising CO_2 emissions could set off rapid global warming."

Vaclav Smil (1997, 2001)

"What are the [global warming] stakes? What are we risking by continuing to play? What do we stand to lose if the worst does come true? Finally, and most importantly, are we so wedded to the choices that have forced us into this game that we cannot bring ourselves to even think about the stakes or to question the risks? Are we like addicted gamblers who can't walk away no matter what it costs to stay in the game?"

Lydia Dotto (1999)

Prudence versus reckless gambling

No sensible homeowner in the modern world would ever say, "I am not going to buy any fire insurance for my house unless and until I see flames licking away at the roof." Similarly, no sensible purchaser of a brand new car would ever say, "I am not going to buy any theft insurance for this car, nor will I even lock its doors, unless and until I see someone trying to steal it." The prudent course of action in these situations is to engage in risk anticipation and risk management. That is what we do when we buy fire insurance, when we wear a personal flotation device while boating, when we refuse to play golf in a lightning storm and

143

when we insist on using a flashlight to find our way along a dark and unfamiliar path at night.

Sometimes, however, prudence escapes us. Sometimes we just cannot bring ourselves to focus on risk anticipation and risk management. Rather than finding a flashlight or hiring a guide or waiting until morning, we set off blindly down the dark footpath, whistling loudly and hoping for the best.

For a striking example of the ease with which prudence can escape us, we need look no further than our current response to the risk—not the certainty but the risk—of human-caused global warming and global climate change. Essentially, that response has amounted to whistling in the dark. We have been telling ourselves that there is no need to do anything significant yet, that true global warming may not even have begun and that it may never begin!

Somehow, we seem to have persuaded ourselves that the risk of significant global warming and global climate change does not fall into the same category as other risks; it therefore requires neither risk anticipation nor risk management. In short, no prudence is necessary. Instead, we tell ourselves, our present behaviour is perfectly reasonable and will continue to be perfectly reasonable at least until scientists come up with incontrovertible proof that human-caused global warming and global climate change are already happening. Somehow, we see no similarity at all between (a) our devil-may-care attitude towards our emissions of greenhouse gases and (b) the behaviour of a homeowner who neglects to buy affordable fire insurance or the behaviour of a golfer who neglects to pay attention to the sky.

Enter Intelligent National Frugality (INF) economics. For any highly industrialized country that chooses to adopt it, INF economics constitutes an effective tool for vastly reducing the amounts of carbon dioxide (CO_2) and other greenhouse gases that the country in question emits into the atmosphere. Indeed, a drastic reduction in a country's consumption of fossil fuels, all of which contain substantial amounts of the element carbon (C), cannot help but result in a corresponding reduction in that same country's emissions of CO_2.

Admittedly, no small-population country such as Canada can have a large direct effect on the risk of global warming and global climate change. The risk is indeed global in scope, requiring cooperative management on a global scale. But there is nothing to prevent even a small-population country, especially one whose present per capita emissions of greenhouse gases surpass those of most other countries, from having the courage of its convictions and from setting an excellent example for others to follow.

The climate-related reasons for adopting INF economics can thus be likened to the normal reasons for buying fire insurance. In both cases, risk gets managed rather than being ignored. Prudence wins out over the urge to gamble. There is, however, an important difference between these two instances of prudence. Fire insurance does not actually lessen the physical likelihood that a particular house will catch fire; it only lessens the likelihood that the homeowner in question will suffer an unaffordable financial loss. On the other hand, INF economics, if widely adopted around the world, would be expected to actually lessen the physical likelihood and/or the physical severity of any future global warming or global climate change.

Accordingly, a decision to adopt INF economics is not as analogous to the purchase of fire insurance as it is to the prudence of the golfer who hastily leaves the golf course as soon as lightning threatens. In this kind of risk management, we deliberately alter our previous or planned behaviour once we realize that circumstances have changed and that new risks confront us. In our personal lives, we are all experts at this kind

145

of risk management, even though nearly all of us do throw caution to the winds from time to time.

To end this section, let me make one further distinction. Some of us, for reasons given throughout this book, would be in favour of INF economics even if the Earth's atmosphere were not at risk. Certain other people, however, while not persuaded by my non-climate-related arguments, might still be prepared to accept INF economics in order to counter the risk of global warming and global climate change. Members of this latter group of people, just like the prudent golfer, are prepared to give up some of today's pleasures in order to have more chance of enjoying tomorrow's. But we in the former group would be giving up nothing that was of particular value to us in the first place. Not being golfing enthusiasts, so to speak, we would not have chosen to spend the day playing golf no matter how cloudless the sky. (My apologies to golfing enthusiasts!)

Relevant scientific certainties

Risks typically involve combinations of the known and the unknown, i.e. combinations of certainties and uncertainties. We do not normally use the word "risk" in situations where the outcome is known in advance with reasonable certainty. We do not normally talk, for example, about the "risk" of being killed if, without a parachute, a person falls out of a flying airplane. Accordingly, this section will look at some key scientific certainties, while a subsequent section will look at some major uncertainties.

In connection with global warming and global climate change, the existence of the so-called "greenhouse effect" constitutes by far the most important scientific certainty. In the words of Ann Henderson-Sellers, "The greenhouse effect is a theory which is

146

well understood by atmospheric and climatic scientists and which successfully predicts temperatures on the Earth and on other planets." No atmospheric scientist in the world, as far as I know, disputes that theory.

Here is a very brief description of the greenhouse effect. The Earth's atmosphere consists of a mixture of gases. By far the most prevalent gas in the mixture is nitrogen gas (N_2), and the second most prevalent is oxygen gas (O_2). Neither of those two fall into the category of greenhouse gases. Mixed in together with the two principal gases are very small quantities of several different trace gases, some but not all of which do fall into the category of greenhouse gases. The main greenhouse gases in our atmosphere are water vapour (H_2O), carbon dioxide (CO_2), methane (CH_4), nitrous oxide (N_2O), ozone (O_3) and certain synthetic gases, including various refrigerant gases.

(Note that the term "nitrous oxide" can be extremely confusing because in some contexts it refers not to N_2O but rather other oxides of nitrogen; this is especially true when the term is used in the plural form: "nitrous oxides". The unambiguous scientific name for N_2O is "dinitrogen oxide".)

Greenhouse gases have been given that name because—in a somewhat similar fashion to the panes of glass in a greenhouse—they allow incoming solar energy (which consists mostly of visible light) to pass through them quite readily, but they interfere with the outgoing passage of the infrared energy that is constantly being radiated outwards from the surface of the Earth. This interference with outgoing infrared energy causes the greenhouse effect.

Infrared energy goes by various other names as well, including infrared radiation, long-wave radiation, thermal radiation and heat radiation. It is perhaps most familiar to us as the warmth-creating energy that radiates outwards from the hot metal surfaces of a wood-burning stove. But relatively cold objects as well, including all parts of the Earth's surface, even the icy parts, emit small amounts of infrared energy.

What happens when the above-mentioned interference occurs is that a portion of the outgoing infrared energy gets absorbed by the greenhouse gases present in the lower atmosphere. Such absorption then warms up both the lower atmosphere and the surface of the Earth. Were it not for this greenhouse effect, the average temperature on the Earth's surface would be about 33 Celsius degrees cooler than is actually the case. In other words, we human beings owe our very existence to the greenhouse effect of the Earth's atmosphere. On the other hand, over on our planetary neighbour Venus, an incomparably stronger greenhouse effect attributable to that planet's dense CO_2 atmosphere has rendered the surface of Venus far too hot to support life.

The greenhouse effect does not actually prevent heat from leaving the Earth. If it did, the Earth would keep on getting hotter and hotter indefinitely. The greenhouse effect simply raises the equilibrium temperature—at the Earth's surface—at which the absorbed incoming energy from the sun is matched in amount by the outgoing infrared energy from the Earth. A long period of time (decades, if not longer) may elapse, however, between the onset of an enhanced greenhouse effect and the attainment of a new, higher, equilibrium temperature at the Earth's surface.

Not all greenhouse gases have an equally strong greenhouse effect. Carbon dioxide (CO_2), for example, has a much weaker greenhouse effect, on a molecule-for-molecule basis, than does methane (CH_4). Nevertheless, carbon dioxide is considered to be the most worrisome of all the greenhouse gases because (a) human beings are currently responsible for enormous emissions of CO_2 into the atmosphere and (b) under conventional economics we are finding it extremely difficult to reduce our CO_2 emissions significantly without turning our economic world upside down.

In addition to the existence of the greenhouse effect, two other key scientific certainties need to be kept in mind. One is that the atmospheric concentrations of three major greenhouse gases in particular, namely carbon dioxide (CO_2), methane (CH_4) and nitrous oxide (N_2O), are all increasing at significant rates and have been doing so for quite some time. Atmospheric concentrations of carbon dioxide, for example, are now (in 2014) in the vicinity of 395 parts per million by volume (ppmv), whereas the figure was approximately 280 ppmv in the middle of the 18th century.

The other key scientific certainty concerns the explanation for most, if not all, of the increase in atmospheric concentrations of carbon dioxide. One does not in fact have to be a scientist in order to see the obvious: The atmosphere's natural ability to rid itself of excess quantities of carbon dioxide cannot keep up with the pace at which the human race is causing new carbon dioxide to be emitted into the atmosphere. We are indeed burning so much fossil fuel and engaging in so much deforestation that unnaturally large amounts of carbon dioxide are being emitted into the atmosphere on an ongoing basis. In addition, we are manufacturing large quantities of Portland cement (the key ingredient in concrete); for reasons of chemistry, as mentioned in Chapter 10, such manufacture gives off substantial quantities of carbon dioxide.

The above three certainties have set the stage for the risk— not the certainty but the risk—that at some point an enhanced greenhouse effect may begin to occur and, once having begun, may last a very long time. (It is difficult to read such books as *Field Notes from a Catastrophe* by Elizabeth Kolbert and the first two chapters of *Eaarth* by Bill McKibben without concluding that an enhanced greenhouse effect has already begun.) The phrase usually used to describe an enhanced greenhouse effect is "global warming". If human causation is involved or under discussion, the warming is often referred to as "anthropogenic" global warming.

Emissions versus concentrations

"Emissions" need to be distinguished from "concentrations". In the case of carbon dioxide (CO_2), for example, we can talk about emissions of carbon dioxide into the atmosphere (so many tons or tonnes per year) and we can also talk about concentrations of carbon dioxide in the atmosphere (so many parts per million as of a particular day or a particular year). The word "emissions" refers to a flow of something from one place to another. The word "concentrations" refers to a quantity of something in a particular place at a particular time (as does the word "abundance").

The greenhouse effect depends upon the concentrations of various greenhouse gases in the atmosphere. But these concentrations generally depend in turn upon the quantities, both past and present, being emitted into the atmosphere. The greater the annual quantity of CO_2, for example, that gets emitted into the atmosphere, the greater will be the concentration of CO_2 in the atmosphere, unless there happens to be a corresponding increase in the rate at which CO_2 is being removed from the atmosphere by such processes as (a) photosynthesis and (b) chemical reaction between basalt rock and any CO_2 that is dissolved in rainwater.

The atmospheric concentrations of two greenhouse gases in particular, however, do not depend upon the emissions of those two gases. The two are water vapour (H_2O) and ozone (O_3).

Water vapour enters the atmosphere (a) when water evaporates from the surface of the oceans and other large bodies of water and also (b) when coal burns, when hydrocarbon fuels (such as gasoline, diesel fuel and natural gas) burn and when trees and wood burn. But water vapour also gets readily removed from the atmosphere in the form of rain and snow and hail. So atmospheric concentrations of water vapour are primarily dependent upon weather and climate. No one spends much time talking about emissions of water vapour, even though they do in fact occur.

As for ozone (O_3), it gets formed in the atmosphere, from oxygen gas (O_2), by means of atmospheric chemistry. Ozone does not get emitted into the atmosphere from somewhere else.

Ozone actually plays different roles in different parts of the atmosphere. Up in the stratosphere, the presence of ozone is entirely natural and serves to minimize the amount of harmful solar ultraviolet radiation that penetrates down to the surface of the Earth. But down in the troposphere, which is the lowest portion of the atmosphere, the atmospheric generation of ozone can for the most part be traced to human activities involving the combustion of fossil fuels. Two complaints can be made about tropospheric ozone: (a) it is a pollutant associated with photochemical smog and (b) because it is a greenhouse gas and because its presence is for the most part unnatural, it may contribute to an enhanced greenhouse effect.

The word "pollutant"

Some people use the term "pollutant" when referring to greenhouse gases. Other people consider that carbon dioxide in particular should not be described as a pollutant at all, since its presence in the atmosphere is absolutely essential if plants are to be able to engage in photosynthesis. In my view, there is no right or wrong here. I myself prefer to describe CO_2 as a greenhouse gas rather than as a pollutant.

Relevant scientific uncertainties

All the scientific uncertainties that surround the risk of global warming boil down to a single uncertainty: We do not know with absolute certainty what, if anything, is going to happen to the global climate and to regional climates as a result of the ongoing increase in the atmospheric concentrations of certain greenhouse gases. We are like the golfer who does not know when, whether or where lightning is going to strike. Like the golfer, we can see that ominous clouds are approaching. But like the golfer, we cannot be certain that harm awaits us.

One possibility is that no harm at all awaits us. That may well be true. But unfortunately, we do not know whether it is true or false.

Another possibility is that great harm awaits us or perhaps awaits our great-grandchildren. That may be true. Again, we do not know. Sea levels may rise if the global climate warms, and low-lying islands and coastal regions may be flooded. Rainfall patterns may be severely altered. Agriculture in some areas may be devastated. Many forests may wither. Tropical diseases and insect pests may extend their range. Away from the tropics, summertime hot spells may increase in both severity and duration. Large-scale human dislocation and conflict may ensue.

Perhaps the main reason why we cannot predict with certainty what will happen if we keep on sending masses of greenhouse gases into the atmosphere is that the question of feedback is exceedingly complex. If negative feedback predominates over positive feedback, then the global climate will presumably continue more or less as before. But if the reverse occurs, if positive feedback predominates over negative feedback, then serious global warming is virtually guaranteed. At present, no one knows for sure which kind of feedback is going to predominate.

Here is an example of negative feedback, i.e. of the kind of feedback that tends to nullify any change that has already begun. Suppose that global warming does begin. Suppose further that the resulting higher temperatures cause higher concentrations of water vapour to be present in the atmosphere. Suppose further that increased cloudiness results. And suppose finally that this increased cloudiness has the result of slightly cooling off the surface of the Earth by increasing the portion of incoming solar energy that simply gets reflected back into outer space. Suppose, in short, that an initial amount of global warming ends up causing an equivalent amount of global cooling. Such a scenario, if it were actually to happen, would constitute an example of negative feedback.

Positive feedback does the exact opposite. It is defined as feedback that tends to accentuate whatever change has already begun. Both ice and snow, for example, reflect a good portion of the incoming solar energy that strikes them; they reflect it back out to outer space. But when rising temperatures cause ice and snow to melt away, thus allowing sunshine to strike the relatively dark-coloured ocean waters or the relatively dark-coloured land (whether treed or treeless), more of the incoming solar energy gets absorbed—and converted into heat—instead of being reflected away harmlessly. Hence, writing in *Scientific American* in November, 2012, John Carey observes: "The [positive] feedback that scares many climate scientists the most is a planetary loss of ice," and "More sea ice around the Arctic Ocean is disappearing than had been forecast."

Similarly, if global warming begins, and if the resulting higher temperatures cause significant quantities of underground (in permafrost) or undersea methane hydrate deposits to release methane (a greenhouse gas) into the atmosphere, and if that extra methane thereupon causes additional global warming, then that too would be an example of positive feedback.

What is worrisome about these and other hypothetical instances of positive feedback is not that they are guaranteed to occur, but rather that no one can guarantee they will not occur. In other words, our present behaviour with respect to emissions of greenhouse gases constitutes a classic example of gambling. Lydia Dotto makes that same point throughout her 1999 book titled *Storm Warning: Gambling with the Climate of Our Planet*. For my part, I consider such gambling to be reckless and selfish. The winnings, if we win, will be puny at best, while the losses, if we lose, may well be irreplaceable and will probably be suffered more by our descendants than ourselves. In the words of James Lovelock, "The most awful thought about what we humans are now doing to the world is the length of time the injuries will take to heal." Carl Sagan made a similar point in 1997: "Thus, even if we stopped all CO_2 and other emissions tomorrow, the greenhouse effects would continue to build at least until the end of the next century. This is a powerful reason to mistrust the 'wait-and-see' approach to the problem—it may be profoundly dangerous."

As many people have pointed out, the human race is at present performing an unprecedented experiment. The experiment asks, "What will happen if we inject large quantities of greenhouse gases into the Earth's atmosphere?" In general, scientific certainty requires that any relevant experiment be able to be repeated again and again, with the same result each time. Clearly, that is not possible in the case at hand. Hence advance scientific certainty regarding the outcome of our current experiment is simply not possible. Wise behaviour, however, is always possible.

Manipulative word games

Some word games get played, either deliberately or subconsciously, for the purpose of deceiving or misleading other people. One such game owes its effectiveness to the easily overlooked fact that the word "evidence" has two different meanings. One of those two meanings is "proof". The other is "facts and observations which, while not amounting to proof, lend significant support to the conclusion that such-and-such a statement is possibly or probably true." From reading crime novels, we are all familiar with this second meaning wherein "evidence" can be rebutted, while "proof" is "proof".

For a typical example of the way in which this "evidence" word game gets played, let us back up several decades to the days before the link between cigarette smoking and cancer was clearly established. To play the game in those days, a person (sometimes an expert, sometimes a non-expert) would state, "There is no evidence that cigarette smoking causes cancer." This person would hope that his or her listeners would interpret those words to mean "There are no known facts or observations which, while not amounting to proof, lend significant support to the idea that

155

smoking might be a cause of cancer," in other words, "There is no reason in the world for thinking that smoking might be a cause of cancer."

In reality, of course, there were plenty of known facts and observations which, while not amounting to proof, lent significant support to the idea in question. But, if challenged, the game-player could always defend himself or herself by claiming, in effect, that the intended meaning of "evidence" was "proof" and that indeed there was as yet no proof that smoking could cause cancer. With luck, no challenge would be forthcoming, the game would be played successfully and the listeners would go away believing that any alleged link between smoking and cancer had no scientific basis whatsoever.

The "evidence" word game is now being widely played in connection with global warming and global climate change. Those who play the game want the rest of us to stop worrying (or not to start worrying) about the global climate. They therefore say, "There is no evidence that ... " and they then add whatever words they think are appropriate. They might say, for example, "There is no evidence that human activity is causing global temperatures to rise." That is actually a very clever statement to put together because it combines (a) the "evidence" word game with (b) the unstated implication that global warming is not something to worry about unless and until it has already started.

It may well be true that as yet there is no proof that human activities are causing global temperatures to rise right now. But that is beside the point. What everyone ought to be acknowledging and pondering is the following: There are plenty of known facts and observations which, while not amounting to proof, lend significant support to the conclusion that today's human activities, if they continue unchanged, are likely to cause future global warming and future global climate change, with unpredictable regional consequences. If you do not believe that, please go back and re-read my section titled "Relevant scientific certainties".

A somewhat related word game, or at least an example of fuzzy thinking, is encapsulated in the following excessively compressed expression of one of the basic principles of modern science: "Nothing is true unless and until its truth has been proven." As a moment's thought confirms, lots of statements must be true even though their truth has not yet been, and might never be, proven. (Surely, for example, one of the following two statements must be true: "Life exists elsewhere in the universe" and "Life does not exist elsewhere in the universe." We just do not know which one of the two is true, and we may never know.) A much more accurate formulation of the principle would be: "Nothing can be counted on as being true unless and until its truth has been proven."

In the case of the global-warming hypothesis, the absence of proof does not mean that the hypothesis is therefore false. Nor does it mean that the hypothesis can safely be ignored unless and until its truth has been proven. Nor does absence of proof tell us anything about probabilities and likelihoods. It simply tells us that we do not know for certain what is going to happen.

Much of what is going to happen in the future is not at all susceptible to foreknowledge. Consequently, we all have to live our lives—and decide upon our behaviour—based on whatever wisdom we are able to come up with.

The nuclear alternative

In recent years, as I mentioned in Chapter 8, nuclear proponents, including newly convinced nuclear proponents, have been quick to point out that nuclear reactors do not emit greenhouse gases. But in my view, the risk-taking involved in the commercial generation of electricity by means of nuclear reactors is every bit as reckless and selfish and unnecessary as the risk-taking involved in the large-scale emission of greenhouse gases. Why choose, as one's escape route from the sizzling frying pan, a path that leads directly to a very hot fire?

Even from the greenhouse-gas standpoint, nuclear power is not quite as pure as the driven snow. As Helen Caldicott has pointed out, "[C]arbon dioxide is emitted at each stage of the nuclear fuel chain, from uranium mining, milling, enrichment, fuel fabrication, construction of the reactor, transportation and storage of radioactive waste, and decommissioning of old reactors. So nuclear power adds to greenhouse warming as well as to radioactive pollution."

By the time a new nuclear reactor has been constructed and tested and is ready to produce its very first megawatt-hour of commercial electricity, it has already acquired a significant greenhouse-gas "debt" as well as a significant energy "debt". Put another way, any new nuclear power station cannot help but make atmospheric matters worse while it is being built. Only after its greenhouse-gas "debt" has been repaid in full—by means of the nuclear generation of a corresponding amount of commercial electricity—can the new station be said to have even begun to help reduce greenhouse-gas emissions. As Andrew Nikiforuk has written, "A nuclear power plant might have to run for a decade before it could legitimately call itself 'free' of carbon dioxide."

Moreover, calculations of the alleged greenhouse-gas benefits of nuclear power are nearly always based on the assumption that we cannot or will not cut back substantially on our overall consumption of electricity. For those of us who reject that assumption, such calculations are totally irrelevant. The less electricity our society manages to consume each year, the weaker the greenhouse-gas argument in favour of nuclear power becomes. For that reason, although I have the greatest respect for most of his views, I disagree strongly with James Lovelock's endorsement of nuclear power in his 2006 book *The Revenge of Gaia*. Neither Lovelock nor anyone else knows just how low we will be able to get our total consumption of non-renewable energy, without giving up anything important, once we tackle the matter seriously.

Carbon sequestration, de-sequestration and re-sequestration

As already mentioned, carbon dioxide (CO_2) is in many respects the most worrisome greenhouse gas in the Earth's atmosphere. For that reason, the topic of carbon (C) sequestration is often raised. If significant quantities of carbon atoms can feasibly be sequestered (i.e. held prisoner, so to speak) in some location other than the atmosphere, then for as long as the sequestration continues, those carbon atoms cannot contribute to atmospheric concentrations of CO_2. In short, no carbon atom can exist in two different places at the same time!

Of course, nature has already sequestered a huge number of carbon atoms in underground deposits of coal, oil and natural gas. Such sequestration, we know, can last for eons. But obviously it comes to an abrupt end whenever human beings extract and burn a fossil fuel and thereby produce CO_2 as an unwanted waste product.

One can use the concept of carbon sequestration to summarize the difference between conventional economics and Intelligent National Frugality (INF) economics. The former tends to shudder at the very thought of deliberately leaving carbon atoms untouched underground in deposits of fossil-fuel energy

159

resources. More precisely, conventional economics offers us a price structure that strongly encourages us, both as a society and as individuals, to build a good part of our lives around the deliberate and massive "de-sequestration" (i.e. release) of carbon atoms. We de-sequester carbon atoms (C) by extracting them from the Earth—in the form of raw fossil-fuel resources—and then combining them with oxygen atoms (O) from the air, thereby forming gaseous molecules of carbon dioxide (CO_2). In contrast, INF economics is designed to enable the carbon in raw fossil-fuel resources to remain untouched far into the future, with only small and ever-decreasing human-caused losses each year.

One can also use the concept of carbon sequestration to point out a major environmental weakness associated with the bitumen deposits found in oil sands, notably the oil sands of northern Alberta. (Oil sands are also called tar sands. For obvious reasons, people in the Canadian oil industry prefer to call them oil sands, while many environmentalists and many non-Canadians call them tar sands. For diplomatic reasons, I use the term "oil sands" in this book.) In a nutshell, very large quantities of other fossil fuels, especially natural gas, get consumed in the process of extracting and refining bitumen, much more so than in the case of conventional deposits of crude oil. This means that when you or I burn any fuel refined from the bitumen in oil sands, we are in effect responsible for the de-sequestration of considerably more carbon atoms than if the fuel had been refined from conventional crude oil (other things being equal). That is why some people are using the term "dirty oil" to refer to what comes out of the oil sands. It is also why a wise Canada and a wise Alberta need to be quite humble about our oil sands and about our present and future plans to exploit their stored chemical energy.

This brings me to "CCS". Given all the carbon de-sequestration that has been taking place around the world in recent decades, many people have proposed that in future we try to re-sequester significant quantities of newly de-sequestered carbon atoms. The term usually used here is "carbon capture and storage" or "CCS".

Conceptually speaking, the simplest re-sequestration idea is to inject large quantities of unwanted CO_2 into empty or partially empty underground storage caverns. Such unwanted CO_2 would generally have come into existence as a waste product of either (a) fossil-fuel combustion in a large power plant or else (b) some other large-scale industrial operation such as the conversion of coal into a gaseous fuel (coal gasification). Note that in underground sequestration of this kind, the carbon atoms get sequestered while still incorporated in molecules of CO_2. No chemical changes are necessary.

Currently, near Weyburn, Saskatchewan, scientists are conducting a major multi-year underground CO_2 sequestration experiment. And off the coast of Norway, for the legitimate purpose of avoiding a carbon tax, a natural-gas company is sequestering CO_2 by injecting it under the seabed. In the Norwegian case, the CO_2 actually originates under the seabed (where, already in the form of CO_2, it is found mixed together with natural gas) rather than being produced chemically in some large-scale industrial operation. That difference is irrelevant, however, from the point of view of the greenhouse effect.

Notwithstanding the greenhouse-effect benefits, there are strong arguments against the deliberate underground sequestration of CO_2. In particular, knowing that CO_2 is capable of causing death by asphyxiation, how much of the stuff do we want to bury under our feet and hope that it stays buried forever? Does it not make more sense, with CO_2 as with virtually all other waste products of our economy, to try to reduce production of the waste in the first place rather than to go ahead and produce it in large and environmentally risky quantities and then try to find a suitable "sink" in which to dump it? It is worth recalling the 1986 tragedy in Cameroon (West Africa) in which various natural forces, having previously caused large quantities of CO_2 to become dissolved in the depths of Lake Nyos, somehow caused that same CO_2 to emerge suddenly in gaseous form from the lake and kill 1,700 people nearby.

Another argument against deliberate underground sequestration of CO_2 relates to energy losses. Such sequestration requires significant expenditures of both money and energy. Why increase our total consumption of energy—by adding the energy expenditures associated with underground sequestration to the previous total—just in order to maintain our current level of energy extravagance? Why not focus instead on frugality and energy efficiency so that no such sequestration is necessary in the first place? After all, quite apart from the question of global warming, there are a great many benefits to be had, both for ourselves and for future generations, if we can bring ourselves to take intelligent frugality seriously.

A different sequestration idea focuses on planting large numbers of trees all around the world. Each year, as they grow, these new trees would extract significant quantities of CO_2 from the atmosphere, since CO_2 is a required feedstock for the biological process of photosynthesis. Carbon atoms would thus be withdrawn from atmospheric molecules of CO_2 and sequestered in the various molecules (principally cellulose, lignin and hemicellulose) that make up the wooden components of trees: wooden roots, wooden trunks and wooden limbs and branches.

Even after a tree dies or is cut down, the sequestration of its carbon atoms will continue until the wood in question either rots or burns. Thus, wooden buildings, wooden boats, wooden docks, wooden boardwalks, wooden furniture, wooden toys and so on, not to mention both standing dead trees and fallen dead trees, all continue to sequester their carbon atoms until rotting or burning finally destroys the wood in question.

There is no argument that I know of against using forests as a means of sequestering carbon. Hence the following suggestion from Dayna N. Scott: "If the [Canadian] federal government insists on linking forests with its climate change strategy, then it should be encouraged to reforest degraded and abused lands, and to protect more old-growth forests as 'national carbon parks' to ensure that these enormous stores of sequestered carbon remain undisturbed."

Obviously, however, the amount of land available for forestation and reforestation is limited, especially in view of our human need for food crops. And even if promoted assiduously, carbon sequestration in wood would not necessarily keep up with all the carbon de-sequestration caused by the combustion of fossil fuels.

In any case, quite apart from carbon sequestration, there are many excellent environmental reasons for planting trees and promoting large healthy forests. So by all means, let us put down most of our chainsaws and pick up our seedlings and our spades. But let us also remember that poor old conventional economics has a terrible time trying to deal sensibly with forest management, just as it does with fishery management, water pollution, topsoil erosion and so on. What happens in all these cases is that conventional economics places a ridiculously low price on the non-renewable energy that can so easily be used to degrade or destroy the resource in question. Simultaneously, conventional economics places a discouragingly high price on the human energy that could otherwise be used generously to safeguard and enhance the resource. Put another way, all roads seem to lead to

Rome, with Rome in this case being represented by INF economics (or something comparable). Whether we want to slow down the de-sequestration of fossil-fuel carbon, or to increase the re-sequestration of carbon in wood, or, ideally, to do both at the same time, Rome seems to me to be beckoning more persuasively than any other destination.

To end this discussion of carbon sequestration and de-sequestration, and to end this chapter, here is a Gaia-inspired perspective from Thomas Berry: "A person can only marvel that scientists generally seem never to have reflected on or explained to the community why the petroleum is buried in the Earth in the first place. Even the slightest reflection would reveal that nature has taken great care to bury the vast amounts of carbon in the coal and petroleum in the depths of the Earth and in the forests so that the chemistry of the atmosphere, the water, and the soil could be worked out with the proper precision. This needs to be thoroughly understood and respected lest anyone intrude into this delicate balance by extracting and using the petroleum or the coal or by cutting down and using the great forests of the planet without consideration of what will happen when these forces will no longer be able to fulfill their role in the integral functioning of the planet."

12

Why Not a Carbon Tax?

"A slower but safer approach [than geoengineering the atmosphere] would be to price greenhouse-gas emissions, preferably through a carbon tax, which would encourage the adoption of cleaner technologies."

The Economist (June 16, 2012)

"Global warming is a product of carbon emissions produced by burning fossil fuels, so, if we want to limit warming, these emissions have to be phased out. Economists on both sides of the political spectrum agree that the most efficient way to reduce emissions is to impose a carbon tax."

Elizabeth Kolbert (2014)

Two different taxation approaches

A great many people around the world are proposing that national governments respond to the greenhouse-gas emissions problem by imposing a carbon tax. A carbon tax would be levied, under certain specified circumstances, either on the chemical element carbon (C) or else on the chemical compound carbon dioxide (CO_2). In either case, the purpose of the tax would be to penalize, and thereby discourage, those human activities that cause excessive amounts of carbon dioxide to get emitted into the atmosphere.

Given the widespread (but by no means universal) support that exists for a carbon tax, the question arises as to whether the Intelligent National Frugality (INF) tax on nuclear and fossil-fuel energy meets the definition of a carbon tax. If not, how and to what extent do the two taxes differ from each other? And which of the two is likely to be more effective in curbing emissions of greenhouse gases? My answers follow.

Based on the words by which it is described, the INF tax clearly falls into the category of an energy tax rather than into that of a carbon tax. Nevertheless, the INF tax will have the effect of penalizing, and thereby discouraging, a large number of human activities that cause carbon dioxide to get emitted into the atmosphere. More precisely, the INF tax will have the effect of raising the purchase price of all fossil-fuel carbon entering the national economy. In that sense, the INF tax can be thought of as constituting one form of a carbon tax.

For reasons of clarity, however, it is probably better to reserve the term "carbon tax" for those taxes that explicitly tax carbon or carbon dioxide. That definition excludes the INF tax. In the remainder of this chapter, I will examine the difference between the INF tax and a true carbon tax. The essence of what I am about to say is that there are significant conceptual differences between the two taxes but as a practical matter those conceptual differences might not amount to very much in the context of what this book is proposing.

Six conceptual differences

Six conceptual differences between the two taxes are worth noting.

One conceptual difference relates to an intrinsic fact about fossil fuels. All fossil fuels contain substantial amounts of the element carbon (C). As a consequence, the combustion of all fossil fuels necessarily produces substantial amounts of carbon dioxide (CO_2). It so happens, however, that one unit of coal

166

energy contains more carbon than one unit of oil energy. And one unit of oil energy contains more carbon than one unit of natural-gas energy. As a matter of logic, therefore, it is impossible to imagine any tax that would (a) tax all fossil-fuel energy at the same rate and (b) simultaneously tax all fossil-fuel carbon at the same rate. A choice must be made. INF economics chooses to tax all fossil-fuel energy at the same rate. Accordingly, the INF tax will have the effect of placing a lower tax on the carbon in oil than on the carbon in natural gas, and a still lower, but by no means negligible, tax on the carbon in coal. A carbon tax, however, would tax all carbon at the same rate.

A second conceptual difference concerns certain emissions of carbon dioxide (CO_2) that are not caused by fossil-fuel combustion. Two instances of such emissions stand out: (a) those produced by the chemical reactions involved in the manufacture of Portland cement and (b) those produced when previously-existing underground deposits of carbon dioxide are released to the atmosphere, notably in cases where such CO_2 is found mixed together with deposits of natural gas (as in the Norwegian situation mentioned in Chapter 11). In both of these instances, the CO_2 emissions lie outside the direct coverage of the INF tax, whereas they lend themselves readily to coverage by a carbon tax. Nothing under INF economics, however, will prevent the government from restricting and/or penalizing the above two kinds of CO_2 emissions in any way it wishes.

A third conceptual difference concerns methane (CH_4), the prime constituent of natural gas. As already mentioned, methane is a much more potent greenhouse gas than is CO_2. A carbon tax, however, would tax the carbon in methane, i.e. the carbon in natural gas, at the same rate as the carbon in oil and coal. Such equality is not found in INF economics. The latter—albeit for reasons relating to energy content rather than to greenhouse potency—has the effect of placing a higher tax on the carbon in natural gas than on the carbon in either of the other two fossil-fuel resources. This is relevant in that the higher tax on natural-gas carbon provides an extra incentive (when compared with a carbon tax) for natural-gas companies and natural-gas pipeline companies to try to prevent any natural gas from escaping into the atmosphere.

A fourth conceptual difference concerns the deliberate underground storage of carbon dioxide as a means of keeping unwanted CO_2 out of the atmosphere. INF economics by itself provides no incentive to consider such storage possibilities, whereas a carbon tax can easily do so. On the other hand, by keeping the fossil-fuel energy tax at a very high level, INF economics may well provide a stronger incentive in favour of frugality and energy efficiency than would any carbon-tax system that rewarded—with a carbon-tax exemption—the underground sequestration of CO_2.

A fifth conceptual difference concerns nuclear energy. Nuclear proponents, including those whose approval of nuclear power is reluctant but real, tend to be delighted with the idea of a carbon tax because the tax would not apply, directly, to nuclear energy. INF economics exhibits no such attraction for nuclear proponents.

A sixth conceptual difference concerns eggs and baskets. Proponents of a carbon tax cannot help but give the impression that they have placed all their eggs in the global-climate basket. In contrast, INF economics has placed its eggs in a variety of different baskets, only one of which relates to the global climate. So even if worries about the global climate were somehow to disappear completely, the attractiveness of INF economics for many of us would hardly be diminished at all.

Two provisos

In spite of all these conceptual differences, I suspect that as a practical matter our society would end up looking much the same under a carbon tax as it would under INF economics, with two very important provisos: The carbon tax would have to be large enough to yield every bit as much revenue as the INF tax, and nuclear power would have to be put out to pasture. I stress the importance of these two provisos, especially since many current proposals for a carbon tax are calling for (a) a much smaller tax than what I am proposing and (b) increased reliance on nuclear power.

The reason for the anticipated similarity of results, subject to the two provisos, is that in each of the two cases there would be very strong financial incentives promoting both frugality and a high degree of energy efficiency. Moreover, neither INF economics nor a carbon tax would have to do everything all by itself. No government is required to tie its hands behind its back and let market forces do whatever tickles their fancy, not even under the wisest of all possible taxation systems. Hence any deficiencies in either system could be dealt with by means of appropriate legislation and regulation.

Why focus on energy?

If the end result is going to be much the same in both cases, why does INF economics feature an energy tax rather than a carbon tax? My general answer is that energy occupies centre stage in economics. (Frederick Soddy pointed this basic fact out years ago. See the Soddy quotation near the beginning of Chapter 1.) Carbon's role, although spectacular at present, is essentially supportive. By focusing primarily on energy rather than primarily on carbon, one can more readily see and understand the many different ways in which our present taxation system is helping to ruin much that we hold dear.

Part 4: Some Familiar Economic Concepts Reconsidered

13

Economic Growth and Its Underlying Ideology

"Notwithstanding the fact that bringing the Jerusalem of economic growth to England's green and pleasant land has so far conspicuously reduced both the greenness and the pleasantness, economic growth remains the most respectable catchword in the current political vocabulary."

E. J. Mishan (1967, 1969)

"But it was only about ten years or so ago that some of us— Georg Borgstrom, Kenneth E. Boulding, Harrison Brown, Ezra J. Mishan, Hugh Nicol, and Joseph J. Spengler to mention only those writers best known to me now—began independently and each one in his own way to challenge the idea of continuous economic growth, which formed at that time (and still forms) the proudest article of standard economic faith."

Nicholas Georgescu-Roegen (1976)

"Hence the first question to ask is whether growth in the economy as measured by GNP actually contributes to the total well-being of people. Until recently, this question was hardly raised, and even today it is not taken seriously in most economic and political circles."

Herman Daly and John B. Cobb, Jr. (1989, 1994)

Growth without end

Let us call a spade a spade. The economic growth so fervently promoted by conventional economics ought more accurately to be called "perpetual economic growth". After all, this alleged need of ours for economic growth does not seem to be based on any specific shortages or insufficiencies. We are simply told that our economy, no matter how large it is this year, will need to be even larger next year and still larger again the following year. Our alleged need for economic growth is insatiable and therefore perpetual.

What about population growth? Does a larger population not justify a larger economy? It may well do so, provided that the economy is sound and non-wasteful in the first place. But population growth does not seem to lie at the heart of the modern world's pursuit of economic growth. Time after time, we read or hear in news reports: "More gloomy economic news today: Growth in the economy this year has barely kept pace with population growth." Under conventional economics, our goal is to ensure annual economic growth over and above whatever population growth takes place.

Perpetual economic growth surely qualifies as an absurd concept. The phrase itself sounds more like a joke or a riddle than like a serious economic proposition. No doubt that is why our culture seldom actually uses the word "perpetual" in reference to economic growth. But that is all the more reason to call this particular ideological spade by its true name.

Our pursuit of perpetual economic growth is the equivalent of saying, "No matter how much I earn today, I will always want to earn even more tomorrow. The same is true of everyone else I know. Under no conceivable circumstances, therefore, should we ever stop striving to make our national economy bigger and bigger and bigger."

Needless to say, we do not often express ourselves quite so bluntly. We prefer to stress that economic growth constitutes a means to an end rather than an end in itself. We need economic growth, we tell ourselves, in order to create additional jobs. In other words, our society may not actually need more goods and services than it is already producing, but it does need more jobs. And without economic growth, we see no way of creating new jobs.

But if conventional economics is truly unable to provide satisfactory numbers of decent jobs in the absence of perpetual economic growth, does that fact itself not constitute a stunning indictment of conventional economics? It is as if a car company were trying to sell a car that kept stalling unless driven faster and faster and faster. A more serious defect in an economic system would be hard to imagine.

As many critics of conventional economics have noted, cancer cells too aim at perpetual growth, until disaster intervenes.

Unlike conventional economics, Intelligent National Frugality (INF) economics does not concern itself with the pursuit of perpetual economic growth, either as an end in itself or as a means of promoting job creation. Instead, INF economics focuses on the direct achievement of legitimate and worthy economic goals: full employment (more precisely, a realistic opportunity for everyone to live a satisfyingly productive life), a healthy environment, frugal consumption of scarce resources (renewable and non-renewable), a reasonable degree of economic fairness and a reasonable standard of living for everyone.

René Dubos cautions us, however: "Normal human beings, it would seem, should find it easy to shift their concern from quantity of production to quality of life. But in practice, the shift will be difficult, because we have been brainwashed into the belief that the betterment of life depends on quantitative growth achieved by an extractive economy."

The costs of economic growth

There is actually a great deal of irony in the importance that our society currently places on perpetual economic growth. Indeed, such growth seems to have long since turned malignant in the advanced industrialized countries of the world. As reflected in Edward Abbey's passionate wit, such growth has endowed his country (the United States) with the world's "grossest national product".

Economic growth of the kind that we have come to consider normal has a great deal to answer for. It has uglified much of our surroundings. It has trampled savagely on the legitimate aspirations of a great many unlucky people. It has contorted our working lives into the proverbial "rat race" in which most of us have to run faster and faster just to remain where we are. It has rained blow after unnecessary blow upon the environment. It has played havoc with our sense of community and our sense of place. And in the view of many people (including myself), it has left our whole society with considerably less wealth than we had before.

Less wealth than we had before? How can a national economy grow larger and larger and at the same time use up more wealth than it creates? Nothing, it turns out, is easier. Modern economic growth feeds off deterioration and depletion and destruction and liquidation far more readily than it nourishes itself with preservation and conservation and truly productive accomplishment. Herman E. Daly and Joshua Farley use the term "uneconomic growth" to encapsulate their observation that "at some point the further growth of the macro-economy could cost us more than it is worth." Daly and Farley have even produced a mathematical type of graph showing conceptually the point at which "economic growth" turns into "uneconomic growth". (I will have more to say about Daly and Farley and their economic views later in this chapter.)

As I see it, the behaviour of modern industrial society is quite similar to that of a young adult who has inherited a very large sum of capital. Unless exceptionally clear-eyed and self-disciplined, such an heir is likely (a) to fritter away all the inherited capital, (b) to do so within a relatively short time and at ever-increasing speed and (c) to hoodwink himself or herself into believing, right up until almost the very end, that the ever-increasing "income" (so-called) being received and spent every year was well and truly and sustainably "earned".

Is that analogy valid? Is the modern industrialized world really confusing "income" with inherited "capital"? All human beings alive today have certainly inherited a huge amount of natural capital from Gaia. (For the meaning of "Gaia", see the Introduction.) As Jane Jacobs' fictional character Hiram Walker observes, "Every [human] settlement starts with at least one useful resource, maybe several, already in place as a gift from nature. A starter resource can be fertile soil or it can be any number of other things: wild animals, flints, nut trees, clay, ore, a waterfall, fossil fuel, hot springs, a beach." Sadly, we moderns have taken to depleting our inherited natural capital at the ever-increasing speed made possible by the combination of ancient nuclear and fossil-fuel energy resources (which themselves form part of our inherited natural capital) and today's technology.

We moderns have also inherited from previous generations a huge amount of social capital in the form of human knowledge, human skills, culture, community spirit, civilized behaviour (albeit

177

with serious imperfections), some degree of stewardship for our particular spot on the Earth and so on. Here too, sadly, we are depleting many aspects of our inherited social capital with unseemly haste and brazen unconcern. In particular—and this is ominous—we are losing our morale; we are beginning to become a demoralized society. As George P. Brockway has written, "[T]hat we shrink from our problems instead of attacking them with eagerness, generosity, and hope is a question of morale."

Just as in the case of the young heir mentioned above, our unceasing economic growth, together with the fantasies that support it, can no doubt continue right up until we belatedly and regretfully discover that we have thoughtlessly squandered far too much of our inherited capital (both natural and social). If I may hark back to the Introduction, we seem determined to kill the goose that has been laying the golden eggs.

Remembering depreciation

Depreciation and modern economic growth are intimately linked. Indeed, the modern world's ever-increasing annual amounts of depreciation have encouraged everyone to believe—falsely—in the legitimacy, feasibility and necessity of perpetual economic growth. Let me explain.

Almost all physical assets produced in whole or in part by human beings have an inevitable tendency to depreciate as time goes by. A portion of their owner's wealth thereby vanishes into thin air. Depreciation is just as real a loss of wealth as would be the case if one dropped some gold coins into the flaming mouth of a volcano.

Not all assets depreciate at the same rate. In each case, the rate depends upon the asset in question and upon the circumstances. Also, the underlying causes of depreciation vary from case to case and do not always include physical deterioration. Examples of possible causes of depreciation include (a) physical deterioration itself (often referred to as "wear and tear"), (b) hygienic deterioration, (c) technological obsolescence, (d) loss of relevant information or documentation, (e) new legislation or new regulations, (f) the whims of fashion and (g) new knowledge relating to human health. My comments on some of those causes follow.

Concerning new knowledge relating to human health, Oliver Sacks sets the stage for an excellent example: "Shoe shops everywhere in my boyhood were equipped with X-ray machines, fluoroscopes, so that one could see how the bones of one's feet were fitting in new shoes." It is easy to imagine the sudden and severe depreciation that such devices must have undergone when they were quietly withdrawn from all shoe stores in response to new awareness of the health hazards associated with exposure to X-rays.

Concerning physical depreciation, an interesting aspect of human nature is that—at least under conventional economics— we keep on forgetting about such depreciation. What home-owner has not discovered, with surprise and dismay, that new roofing has suddenly become necessary because of an unexpected roof leak? What car-owner has not pulled a long face upon learning that new brakes or new tires or new exhaust components must be purchased immediately? In most cases, these are nothing more than predictable matters of depreciation. And yet, how we love to treat them as unexpected bolts from the blue!

This same half-real, half-feigned surprise at physical depreciation can also be found in governments. Depreciation of our government-owned physical infrastructure constitutes a major component of our total national consumption of wealth, and yet somehow our various levels of government are never quite prepared for it. Hence the term "concrete deficit". That term refers to the situation—quite common under conventional economics—where the general state of repair of a great many bridges, highways, piers, sewers, tunnels, etc. (made of concrete to a large extent) is deteriorating significantly because too little maintenance and repair work is being done. For the most part, all this depreciation is perfectly normal and predictable.

In the words of Professor Saeed Mirza, referring to Montreal's municipal infrastructure in particular, "I've been preaching for years that maintenance must not be deferred ... We have to get away from the present philosophy of design, build and forget." Note the word "forget" in that last sentence.

Why is it that conventional economics performs so badly with regard to depreciation? There is a general answer, I think, and also a specific answer.

The general answer is that under conventional economics, with its emphasis on always producing more and more and more, we all end up getting seduced by the attractions of whatever is brand new: a brand new house, a brand new car, a brand new bridge, a brand new kitchen, a brand new pair of skis, and so on. Why else do manufacturers and purchasers of new cars spend so much money on flashy paint jobs and shiny finishes? We have become so enamoured of things that are brand new that we have lost almost all interest in middle-aged goods, middle-aged buildings and middle-aged infrastructure. Even worse, our incipient demoralization is causing us to have less and less concern for the medium-term future and zero concern for the long-term future. "I'll be dead by then," we often hear ourselves saying, as if—for us—nothing that happens after our last breath could conceivably have any importance.

My more specific answer is that conventional economics has a high propensity to produce things that depreciate rapidly. This is as true for infrastructure assets as it is for buildings and consumer goods. And there is no mystery as to why this should be the case. After all, seen from the point of view of INF economics, conventional economics heavily subsidizes most of the energy consumed in our manufacturing and construction and transportation activities. As a result, neither producers nor consumers have much financial incentive to get maximum value—maximum energy value as distinct from maximum money value—out of all that energy. More specifically, we have little financial incentive to maximize the life expectancy of what we produce and little financial incentive to promote any kind of national frugality and overall energy efficiency.

The latter two specific points translate directly into depreciation factors. A short life expectancy has essentially the same meaning here as a rapid rate of depreciation. And weakness in the area of national frugality and in the area of overall energy efficiency implies an unnecessarily large inventory of depreciating assets. When a rapid rate of depreciation is combined with an excessively large inventory of depreciating assets, the overall size of our annual national depreciation becomes enormous. And it keeps getting larger and larger because our total inventory of rapidly depreciating assets keeps getting larger and larger. Put into numerical terms, if our country's annual depreciation losses increase by one billion dollars every year, then just to keep our overall level of national wealth from falling, we have to increase our annual production of wealth by one billion dollars every year. Talk about having to run faster and faster in order to stay in the same place!

As a spectacular example of a rapid rate of depreciation combined with a huge inventory of depreciating assets, consider our current approach to ground transportation. Under conventional economics, we do two things: We give priority to road transportation over rail transportation, and we tend to maximize our overall amount of transportation activity (as regards both freight and people).

In practical terms, this means first of all that we give priority to a system of transportation for which the rates of depreciation—for highways as well as for motor vehicles—are extremely high. Highway surfaces deteriorate very quickly, as we all know, especially when travelled over by a high volume of heavy trucks. Highway bridges too need frequent inspection and repair.

As for motor vehicles, they have notoriously short working lives as currently designed, especially passenger cars and small trucks. Their short lives are shortened even further in those locations where salt is routinely spread over road surfaces for the purpose of melting winter ice. Moreover, as already mentioned, certain major vehicle components such as tires and exhaust systems depreciate even faster than the basic vehicles themselves. Nor should we overlook the sudden depreciation caused by our far-too-frequent motor vehicle collisions.

Secondly, we keep expanding our demand for transportation services, which means that we keep pouring into our main transportation system vast quantities of additional physical assets: additional kilometres of new highway and of widened highway, additional highway bridges and interchanges, and additional motor vehicles. In other words, we keep expanding our inventory of assets that are subject to rapid depreciation.

Sometimes, we compound the depreciation problem still further by deliberately demolishing a perfectly good highway bridge in order to replace it with a brand new and much larger bridge. The value of the old bridge depreciates suddenly to zero— far faster than would have been the case under normal usage—in a scenario intended to accommodate an enlarged inventory of assets (motor vehicles) that are themselves subject to rapid depreciation!

My point, of course, is not that we should eliminate the totality of our transportation assets and thus eliminate all depreciation related thereto. Rather, my point is that, with regard to our transportation assets, conventional economics tends to saddle our society with an increasingly high depreciation bill every year and it also tends to seduce us into overlooking a great deal of that depreciation when we make our basic economic choices.

Transportation provides by no means the only example of the enormous amounts of depreciation that occur under conventional economics. Think of all the buildings that we seem to find it necessary or appropriate to tear down. Think of all the rusting machinery, the broken furniture, the discarded appliances, the obsolete electronic equipment and so on. Admittedly, such items may well have scrap value, but only to the extent of a small fraction of their original value.

Goods intended to be used only once should not be forgotten either. A paper cup generally loses 100% of its value after a single use. The same is true of a bullet, a stick of dynamite, plastic garbage bags and many components in space rocketry.

Still another kind of ongoing loss is perhaps the easiest of all to overlook: the rapid depletion of our total reserves of oil and coal and natural gas. To prevent any misunderstanding, let me be clear that by "total reserves" I mean all the not-yet-discovered reserves as well as all the ones already discovered but not yet exploited. That total, the real total, is moving far too rapidly downwards. Every two weeks, for example, the human race at present is obliterating a billion or so barrels of crude oil. Moreover, the cost—both in money and in energy—of extracting a barrel of crude oil from the Earth keeps increasing as we move from the easy pickings of the "low-hanging fruit" to the difficulties associated with unconventional sources of petroleum, such as the Alberta oil sands and deep undersea deposits, a point that I will be discussing further in Chapter 17.

If we truly cared about keeping meaningful national accounts, we would total up our gross wealth and gross production figures each year and then subtract a comprehensive figure for Gross National Depreciation, Depletion and Destruction.

Actually, it is not the numbers themselves that really matter here. Numbers always need to be treated somewhat skeptically because of the garbage-in-garbage-out problem: A single major inaccuracy or omission or a single unwarranted assumption can invalidate pages and pages of sophisticated calculations or their computerized equivalent. That is why so many critics have expressed dissatisfaction with current methods of calculating "gross domestic product" or "GDP".

Rather than worrying too much about actual numbers, we need to take the time and trouble to look around and see with our own eyes and judge with our own minds what is actually going on and what choices are available to us. If we wish, we can build a productive and non-wasteful economy in which the overall total of depreciation losses and depletion losses is far smaller than at present. That by itself will enable us to lower our overall level of gross production substantially. It will also help all of us to step aside from today's "rat race" and to live and work at a much more reasonable and satisfying pace.

We need to keep in mind the elementary arithmetical fact that—both as individuals and as a whole society—we can constantly increase our net wealth (if we so desire) even in the complete absence of any economic growth or growth in personal income. To do so, we simply need to ensure that each year our production of wealth exceeds our consumption of wealth. We need to remember, however, that our consumption of wealth includes all the disappearances caused by that unobtrusive pickpocket known as depreciation, a pickpocket that has been having unprecedented success under conventional economics.

INF economics, I submit, will wring a great deal of depreciation and depletion out of our economy altogether, including most of the portion associated (a) with frivolous goods, including frivolously short-lived goods, (b) with energy-inefficient infrastructure and (c) with our current energy-inefficient approach to the production of most of our goods and services.

Remembering inequality

There is still another way in which economic growth can lower, or at least fail to improve, most people's material living standards. Here too, just as in the case of depreciation, what is involved is nothing more than simple, albeit easily overlooked, arithmetic.

Imagine a very privileged class of eighteen students. To this lucky class are being forwarded substantial distributions of cash every month. Three of the students, however, happen to be bullies. These three bullies make a point of allocating almost exclusively to themselves all the incoming cash distributions, even though the latter are intended for the class as a whole. Moreover, from time to time the three bullies further victimize their classmates through acts of extortion and petty larceny. As a result, the three bullies become quite wealthy whereas the other fifteen acquire nothing. The average wealth of each student goes up, since the cash distributions are taking place on a regular basis and since the bullies save a portion of their monthly takings. But for fifteen of the eighteen students, there are no benefits at all.

The word "average" is always a crucial word. It conveys the information that what is being discussed is an arithmetical relationship connecting certain facts. The basic facts themselves are not necessarily discussed or even stated.

We all tend to get easily fooled by the word "average". We do so because sometimes "the average person" means the same thing as "most people" and sometimes it does not. In the case of human life-expectancy statistics, for example, the statement that the average life expectancy at birth in country X is 80 years tells us (since we all have a general awareness of the biological length of the human life span) that most people in country X die within a few years of their 80th birthday. But the statement that the average life expectancy in country Y is only 50 years does not tell us that most people there die within a few years of their 50th birthday. In all probability, very few people there die anywhere near the age of 50; but because of high infant mortality, the average age at death is only 50 years. That would be the case, for example, if 70% of the people died at the age of 71 years while the remaining 30% of the people died at the age of 1 year.

In the case of the eighteen classmates, we might be told simply that their average wealth keeps going up. Although perfectly correct, that information would be highly misleading because it would conceal the important fact that the fortunes of the three bullies were so different from those of the other fifteen.

My point here is not to accuse the wealthiest one-sixth of our society of being bullies and extortionists and larcenists. But arithmetically speaking, there might not be much difference between what is actually happening in much of the modern world and what happens among the eighteen classmates.

An editorial in *The New York Times* of February 1st, 2014, referring to the American economy, said much the same thing but in more diplomatic language: "Nor is there any guarantee that the benefits of stronger growth, when and if it materializes, will be broadly shared; to date, what growth there has been has largely benefited those at the top of the income and wealth ladder, a dynamic that becomes more entrenched the longer it endures." And in 2004, Jack Layton wrote the following: "Since 1995, we've seen the most rapid increase in the inequality of incomes among Canadians since we've kept track of these things, and this at a time when the economy has been growing significantly."

Economic growth concerns the overall economy. It tells us nothing about individuals. It tells us nothing about this half or that half of the population. It tells us nothing about this sixth or that sixth. By the same token, the absence of economic growth— or even an actual shrinking of the economy—would also tell us nothing about individuals. The poorest 5% of our population, for example, might well have access to far better food and far better housing under Intelligent National Frugality than they do now under conventional economics.

In that regard, note the following comment by Naomi Klein: "[T]hough overall consumption in the U.K. dropped by 16% [during World War II], caloric intake for the poor increased ... because the rations provided low-income people with more than they could otherwise afford." Klein was making the point that overall frugality (my word, not hers) can, and should, be deliberately combined with a significant improvement in the indecently low living standards of the poorest in our own communities, as we in the wealthy countries of the world face up to the challenge of global warming and climate change.

As I see it, INF economics seems likely to lessen economic inequality substantially. It might even allow us to approach a goal described approvingly several years ago by E. F. Schumacher: that those at the top should receive no more than seven times as much remuneration as those at the bottom. In a great many large organizations at the present time, we are nowhere near that small a ratio.

Three factors in particular are worth noting here. One is that decent jobs and other decent work opportunities will become plentiful under INF economics, and hence the kind of poverty associated with involuntary unproductiveness will tend to disappear. A second factor is that the new tax on fossil-fuel energy, having been converted into a progressive tax by means of the payments discussed in Chapter 4, will tend to extract considerable wealth from the wealthy for the benefit of society as a whole. And the third factor, previously mentioned, is that great concentrations of wealth are probably much easier to accumulate (for entrepreneurs, business and banking executives, entertainers, professional athletes and so on) when great concentrations of energy are readily available than when concentrated energy is very expensive.

I should also mention a fourth factor, a factor that might prove to be the most important of all. Not having as one of its stated goals the pursuit of perpetual economic growth, INF economics will not be tempted to follow conventional economics in trying to divert people's attention away from the topic of economic inequality and onto the topic of economic growth. Conventional economics has been saying to everybody, in effect, "Don't worry about how much richer the rich are than yourself. Just focus on the fact that you'll have more money next year than this year. And more still the year after next. Be thankful!"

According to the values underlying Intelligent National Frugality, economic fairness—which precludes the excessive degree of inequality so prevalent today—constitutes an important goal in its own right. Hence INF economics will invite everyone to judge its performance partly on the basis of its success in keeping economic inequality within reasonable limits.

Let me state again that nothing in INF economics will prevent the government from imposing inheritance taxes, wealth taxes and so on, nor from establishing economic safety nets for the unfortunate. In a democracy, the degree of inequality of wealth need never exceed whatever degree is considered acceptable by a majority of the people.

Ideologies

(Throughout this book, whenever I use the word "ideology" I do so in its narrow pejorative sense. Admittedly, the dictionary meaning of "ideology" is broad enough to include even such positive political concepts as "democracy". But over the last hundred years or so, the world has been shaken by so many negative and hatred-inspired ideologies that the word "ideology" itself has slowly come to denote, in many contexts, any negative and excessively rigid and more or less intolerant political or economic creed. Moreover, no other English word that I know of conveys quite the same flavour with quite the same punch as the word "ideology" when the latter is used in its narrow pejorative sense.)

Personally, I have no use for ideologies at all. They tend to be dishonest, manipulative, hard-hearted, intolerant, ungenerous, intellectually feeble and in the long run socially disastrous. The whole world can now see how strongly those traits characterized the ideology of Soviet communism, the ideology of Hitler's Nazism and the ideology of South Africa's apartheid.

The ideology underlying our current pursuit of perpetual economic growth does not hold a candle, as regards downright evil, to the three examples just mentioned. Nevertheless, it does possess all the attributes of an ideology. What makes it relatively benign, but only relatively, is the fact that we ourselves have freely and willingly chosen to impose this ideological straitjacket over our own minds. Moreover, since we live in a democracy, we are at liberty to remove the straitjacket whenever we wish.

The essence of a political or economic ideology consists of a strong and unhealthy social pressure that tries to promote a narrow conformism and to discourage independent thinking. In the unforgettable phrase of Alfred, Lord Tennyson, "Theirs not to reason why ... " No ideology wants anyone to reason why or to exercise independent judgment, at least not on matters of importance to the ideology in question. As Thomas Homer-Dixon has gently written, "Sometimes the institutional structure of a society or its culture of fundamental beliefs—such as its religion or its political and economic ideology—constrain the society so that it can't properly explore its fitness landscape." Jane Jacob's fictional character Armbruster is more vehement: "First, beware of drift into ideology. Economic ideologies are a curse."

The ideological nature of our society's pursuit of perpetual economic growth is often revealed unwittingly by the media. In theory, for example, our newspapers separate their news stories from their editorial comment. But because of the strong ideological pressure that pervades our whole society, no Canadian newspaper that I know of can resist throwing a congratulatory word or two into just about every news story reporting another three-month period of supposedly "healthy" economic growth. Even the Canadian Broadcasting Corporation does this. And, presumably, so do most of the news-reporting organs around the world. The ideology of perpetual economic growth seems to have all of us in its grip.

Not surprisingly, then, we have all been slow to realize that conventional economics has brought together two mutually antagonistic bedfellows. One, who would much prefer to be sleeping alone, can be described as the free-spirited and non-ideological study of human economic behaviour and human

190

economic institutions. The other, trying very hard to dominate without being clearly identified, is none other than the ideology of perpetual economic growth, masquerading as some sort of economic science.

My criticisms of conventional economics are directed exclusively against the second bedfellow, not the first. After all, no science worthy of the name goes around telling people what political or economic choices to make. No science worthy of the name goes around telling people not to ask awkward and unwelcome questions. No science worthy of the name goes around insisting that economic growth is inherently desirable and that the matter is not open to any question or discussion whatsoever. Such behaviour is typical of the behaviour of an ideology and completely at odds with the behaviour of a science.

Ideally, the role of a non-ideological economic science ought to be to tell society how the economy actually works, to outline the economic choices that are available and to anticipate and explain the probable consequences of each of the various choices. Economic decision-making, however, is essentially a political matter rather than a scientific matter and hence should be treated accordingly. Most important of all, economic science should always strive to maximize the number of economic choices that receive serious consideration. Ideologies do the opposite.

I am not alone in focusing on the ideological nature of our current attitude towards perpetual economic growth. In his book *Something New Under the Sun: An Environmental History of the Twentieth-Century World*, written in the year 2000, historian J. R. McNeill makes the following observations: "Economic growth became the indispensable ideology of the state nearly everywhere ... Like an exotic intruder invading disturbed ecosystems, the growth fetish colonized ideological fields around the world after the dislocations of the [Great] Depression [of the 1930s]: it was

191

the intellectual equivalent of the European rabbit." To the three epigraphs that I have placed at the beginning of this chapter, McNeill offers a disheartening sequel: "The true heretics, economists who challenged the fundamental goal of growth and sought to recognize value in ecosystem services, remained outside the pale to the end of the century."

One's discouragement need not be total, however. "Heretical" economists do indeed exist and are beginning to be heard and even to receive prestigious awards. Herman E. Daly of the United States is probably the best-known and most honoured example in the English-speaking world, even though most books on economics are careful to mention neither his name nor his ideas. For many years now, Daly, sometimes by himself and sometimes with others, has been writing and editing books and articles that further develop the ideas of "ecological economics". His major work, written together with John B. Cobb, Jr. in 1989 and revised in 1994, is titled *For the Common Good: Redirecting the Economy toward Community, the Environment, and a Sustainable Future.* One can see from the title why mainstream economics shuns economists of Daly's ilk!

In 2004, Daly and Joshua Farley together wrote an actual textbook titled *Ecological Economics.* In that book, they quote J. R. McNeill at much greater length than I have done, they call for "a return to the beginnings of economics as a moral philosophy explicitly directed toward raising the quality of life of this and future generations," and they state: "Alternatives to our misguided goal of infinite growth and limitless material consumption will be discussed throughout this text."

Incidentally, Daly and Farley begin *Ecological Economics* with a two-page "Note to Instructors" which includes the following sentences: "We the authors are both economists trained in the standard neoclassical Ph.D. programs that one finds in nearly all

American universities. Between us we have taught and practiced economics for over 50 years in universities and development institutions in various countries ... We accept more of traditional economics than we reject, although we certainly do reject some of the things we were taught."

To end this section and this chapter, I would like to suggest that INF economics does not qualify as an ideology at all, not in the pejorative sense in which I am using that term in this book. INF economics does not claim to be the only possible system that economic science allows. It does not claim that its ideas somehow have greater scientific validity than those of all other approaches. It does not even take sides in the debate between socialism and free enterprise (see Chapter 19). And it certainly does not seek to discourage the asking of awkward questions. Instead, it offers itself as an option, as a choice. "I am available, if you want me", it says. "I believe that I can help translate certain specified human values into social and economic reality. If you share those values, then you may wish to give serious consideration to my basic approach to economics."

14

Sustainability and Frugality

"An ecological economy would measure the prosperity of a society not in terms of the numbers of goods produced, but rather in terms of production methods that conserve the environment, protect human health, and result in durable consumer goods. The measure of value would include clear air, pure water, unpoisoned food, and the flourishing of diverse life forms."

Petra K. Kelly (1994)

"We are unlikely to achieve anything close to sustainability in any area unless we work for the broader goal of becoming native in the modern world, and that means becoming native to our places in a coherent community that is in turn embedded in the ecological realities of its surrounding landscape."

Wes Jackson (1994, 1996)

Overview

Most economic activity in the modern world is light-years away from both sustainability and frugality. Modern factories, modern farms, modern transportation systems, modern mineral-extraction sites, modern nuclear and fossil-fuel power plants and so on tend to be fairly bursting with unsustainability. For future generations, that does not bode well. Caution, however, is required here. Aiming too high can sometimes be almost as undesirable as aiming too low. Nevertheless, we can do far better in this area than at present, even in situations where complete sustainability is either difficult or downright impossible.

Sustainability

If they are going to be used at all in the context of economics, the words "sustainable" and "sustainability" need to be used with great care. A key problem is that all fossil-fuel energy resources are non-renewable. In other words, the use of any fossil-fuel energy resource is inherently unsustainable. How then can one speak of economic sustainability unless one is indeed advocating a complete cessation of the use of all fossil fuels? Not even Intelligent National Frugality (INF) economics goes that far.

What INF economics will do is permit and encourage a vast reduction in our consumption of fossil-fuel energy resources. In fact, I foresee under INF economics an ongoing annual decrease in our consumption of oil and coal and natural gas. As a result, our consumption of those resources will be enabled to continue far longer into the future than would otherwise be the case. The less we consume unsustainably each year, the longer we can continue to consume unsustainably! And when the end finally does come, the weaning process will have been so gradual that the end of the fossil-fuel age will hardly be noticed at all. That, at least, is the goal at which INF economics aims.

By contrast, conventional economics seems quite oblivious to the unsustainability inherent in fossil-fuel energy consumption. Our society is therefore currently running the risk that the transition from high fossil-fuel consumption to low fossil-fuel consumption may end up being disruptively sudden, as many people have been warning. (See the first two quotations at the beginning of Chapter 7.)

A metaphor

If I may use a metaphor here, our present situation features a speeding bus and a long transcontinental thoroughfare. Only for the initial portion of its very long length can this thoroughfare be described as a modern paved highway. And right now, we are driving our fully loaded bus at very high speed along the initial paved portion. This paved portion represents the modern era, i.e. the period of time during which energy has been abundant, inexpensive and extravagantly consumed.

Somewhere up ahead of us, the paved portion ends. From that point on, we know, we will find ourselves on a dirt-surfaced rural road that is not at all suited to high-speed driving. This rural road represents a future era in which energy will be of limited availability and expensive to obtain in large concentrations.

The bus itself represents the behavior of the modern industrial world. The current high rate of speed of the bus represents the current high rate at which we are consuming non-renewable energy.

Whizzing along comfortably in our speeding bus, we do not know exactly how far ahead of us the changeover from modern pavement to a rural dirt surface is located. But since a certain length of paved highway is still visible in front of us, we feel confident that we have enough time to decelerate gradually and safely, if and when we so wish. And provided that we lighten the pressure of our foot on the accelerator pedal immediately, we probably do have enough time.

"Immediately" might turn out to be sooner than necessary. On the other hand, any losses occasioned by our slowing down sooner than necessary would be as nothing compared with the costs and risks involved if we were to continue driving at very high speed and then had to panic-brake at the last moment when the end of the pavement finally came into view. (For two book-length essays on what that panic-braking and its aftermath might look like in the United States, see *The Long Emergency*, written in 2005, and its 2012 sequel *Too Much Magic*, both by James Howard Kunstler.)

Incidentally, my highway metaphor simply embellishes the theory of "peak oil". According to the theory of peak oil, the problem we human beings face is not that we will soon run completely out of crude oil. Rather, the problem is that, admittedly with periodic ups and downs (some of which are quite severe), crude oil is becoming increasingly difficult to obtain and therefore increasingly expensive. At some point, even with the very latest technology, we will never again be able to extract and consume as much oil in the current year as we did in some previous past year. Hence, "peak oil". And the greater the rate at which we consume oil today, the sooner peak oil is likely to be upon us and the more difficulty we are likely to have with what comes next. Actually, all these points about peak oil seem rather obvious, do they not? One wonders why the word "theory" has to be used here at all, given that crude oil is a non-renewable resource.

In the context of raw fossil-fuel resources, the word "production" is a widely used euphemism for "extraction". "Production" suggests renewability, whereas "extraction" implies non-renewability. How we all like to kid ourselves!

As for the end of the petroleum era, it will come long before all crude oil has been extracted from the Earth. It will come, at the very latest, when the oil industry can no longer extract more than one barrel of oil in return for every barrel of oil invested in the industry's operations. Why (indirectly) sink a barrel of "old" oil into the ground in order to extract no more than a barrel of "new" oil? Who would ever finance such a pointless investment? (For more on this point, see the section on EROEI in Chapter 17.)

I should also acknowledge that my highway metaphor totally ignores the threat of global warming and climate change. That threat overshadows the implications of fossil-fuel non-renewability that are under discussion in this chapter. Indeed, even if the thoroughfare in question were known to be paved for hundreds of kilometres in front of us, we would still be wise to begin reducing the speed of our bus immediately. The global warming threat by itself provides us with an urgent reason for doing so, as do many of the other problems discussed in this book. (For my discussion of global warming and climate change, see Chapter 11.)

The transitional nature of INF economics

Given the inherent unsustainability of fossil-fuel energy consumption, INF economics can never be more than a transitional economic system. It can only bridge the huge gap between today's energy-guzzling extravagance and some future economic regime based exclusively on sustainable sources of energy. With luck, however, INF economics might endure for a very long time, despite its inherently transitional nature. The very disincentive that it creates, namely the disincentive to consume large quantities of fossil-fuel energy, should serve to prolong its applicability far into the future, barring the unforeseen.

INF's successor

What taxation system will eventually succeed INF economics? We can safely leave that decision to our descendants. They might end up choosing to tax personal income once again, there being no longer any fossil-fuel energy commercially available to degrade the environment and steal good jobs away from human beings. Or they might end up adopting Henry George's nineteenth-century proposal for a single tax on land. Or they might end up trying something that we have not yet imagined. In any case, our generation need only concern itself with planning and carrying out the transition from conventional economics to INF economics (or to something comparable).

Renewable resources

It is easy enough to understand that the consumption of fossil-fuel energy resources is inherently unsustainable, but what about renewable resources? Forests and fisheries, for example, are quite capable of yielding a sustainable harvest of lumber, firewood, pulp, fish, crabs, lobsters and so on, provided that the resources in question are neither over-exploited nor environmentally harmed. So sustainable logging, sustainable fishing, sustainable farming and sustainable hunting are all perfectly possible, at least in theory.

But there is a fly in the ointment. In our present-day exploitation of renewable resources, we nearly always consume, directly or indirectly, significant quantities of fossil-fuel energy, plus in many cases electricity generated from nuclear energy. In other words, there is an element of unsustainability even in our dealings with resources that are classified as sustainable.

A language problem thus becomes apparent. Does the term "sustainable logging", for example, refer exclusively to the sustainability of the forest itself? Or do the fuels used in the chainsaws and skidders and logging trucks have to be derived from sustainable resources as well? Can the term "sustainable fishing" be applied where fish stocks are wisely managed but the fishing vessels are powered by fossil fuels? And what about "sustainable farming"? If any fossil fuel is used in a farm tractor, does that automatically mean that the whole farming operation should be categorized as "unsustainable"? Otherwise, just exactly where is the point at which sustainable farming becomes unsustainable farming?

My general answer to all these questions would be that the use of fossil-fuel resources should never be allowed to injure in any significant way, either directly or indirectly, the sustainability of renewable resources. Whenever such injury does occur, unsustainability has clearly set in. But in the absence of such injury, the use of such terms as "sustainable logging" and "sustainable agriculture" would seem justified, provided that the intended meaning is understood by everyone.

As always, however, the real world consists mostly of shades of gray rather than of pure black and pure white. In particular, even if no fossil-fuels are used at all, truly sustainable agriculture is not easily accomplished. Precious topsoil can get blown or washed away in excessive quantities unless great care is taken. And the recycling of needed plant nutrients, notably nitrogen (N), phosphorus (P) and potassium (K), requires a great deal of social coordination and knowledgeable effort if sustainability is to be achieved. By itself, INF economics will not guarantee agricultural sustainability. But it will provide a set of financial incentives— and, I hope, an overall social environment—that will encourage us all, farmers and non-farmers alike, to set much higher standards for ourselves in this area than is the case at present.

Sustainable growth and sustainable development

The trouble with the terms "sustainable growth" and "sustainable development" is that they mean different things to different people. To one person, "sustainable growth" may designate a growth process that goes on forever, in other words a kind of perpetual economic growth. But to another person, the term may simply designate any past growth that leads to present stability. Mark Dowie offers a similar comment on "sustainable development": "To a corporate economist sustainable development means development that will allow his company to remain in business forever. To an environmentalist it is development that will allow the earth to stay in business forever. To a deep ecologist it is virtually no development at all as we know it."

There is not even agreement on the basic question of whether "growth" and "development" mean the same thing.

I realize that many people use in good faith either the term "sustainable growth" or the term "sustainable development", or perhaps even both terms, as a way of expressing their disapproval of the modern world's apparent determination to seek short-term gain at whatever cost. But unfortunately, that very determination leads our society to play word-game after word-game with those two terms. My own preference, therefore, is to try to avoid using either term if at all possible.

Matter, energy and sustainability

My final comments on sustainability relate to a point put forward by Nicholas Georgescu-Roegen, one of the twentieth century's most accomplished and most scientifically literate economists. The essence of his point is that in the long run our use of iron (Fe), copper (Cu), aluminum (Al) and other kinds of mineral matter is inherently unsustainable. Matter, Georgescu-Roegen felt, deserves every bit as much attention as does energy. Moreover, from the standpoint of the human race, the connection between *matter* and unsustainability is ultimately much closer than that between *energy* and unsustainability.

Admittedly, fossil-fuel energy epitomizes unsustainability. But solar energy epitomizes sustainability. Solar energy will be arriving on Earth almost forever. And once the human race has weaned itself—either suddenly and painfully or (preferably) gradually and cooperatively—off fossil-fuel energy, there will no longer be a direct connection between (a) the principal form of energy used by human beings and (b) unsustainability. Thereafter, unsustainability will be associated primarily with matter rather than with energy.

Georgescu-Roegen summarized his point as follows: "[I]t is not the sun's finite stock of energy that sets a limit to how long the human species may survive. Instead, it is the meager stock of the earth's resources that constitutes the crucial scarcity."

The problem with mineral matter is that the stuff is of no economic use to human beings unless it is available in at least minimum concentrations. An atom of iron in the soil here and another atom of iron in the soil over there are of no economic use to anyone. But an existing concentration of iron atoms in one place can be very useful indeed.

Within limits, energy can be used to concentrate iron (Fe) or gold (Au) or tin (Sn) or some other mineral. But limits do exist. If the initial concentration is too low, then even a very large expenditure of energy will be fruitless; the loss (de-concentration) of mineral matter through wear and tear and so on will be just as great as any gain (concentration) of mineral matter achieved by the technology being employed.

"Wear and tear" is a key concept here. As the wheel bearings of a car gradually wear out, for example, tiny quantities of metal more or less disappear. The atoms of iron and of other elements do not literally cease to exist, but they do drift farther and farther away from each other in the form of bits of dust and so on. In a word, they become totally de-concentrated. For all practical purposes, they are lost to the human race. The main body of the wheel bearing may well remain intact, but a small percentage of its mass will have flown the coop.

Since wear and tear are inevitable, the gradual de-concentration of mineral matter is also inevitable in any society that uses such matter.

One's initial reaction to this line of thinking tends to be that it is more or less on a par with the information that the sun will one day get too hot for all life on this planet. But there are two important differences. One is that we human beings do not have any say in how hot the sun becomes. The other is that the time-frames involved are not even remotely similar. The scale of our present-day production and consumption of metallic goods is so enormous that major shortages of concentrated mineral matter may occur not too many generations from now.

I have no quarrel at all with Georgescu-Roegen's analysis of this problem. As far as I can see, it further reinforces all the other arguments that underpin INF economics. By far the simplest way of ensuring that *mineral matter* gets used wisely and sparingly is to ensure that *energy* gets used wisely and sparingly. After all, iron ore does not jump out of the ground and form itself into an automobile or a ship all by itself. A great deal of help in the form of energy is required. Our extravagant use of mineral matter during the past hundred years or so could never have occurred without an equally extravagant use of energy. By the same token, any future shortages of concentrated mineral matter could have been postponed for very long periods of time—and to a lesser extent could still be postponed—by a timely decision on our part to begin using energy wisely and sparingly.

INF economics

All "go to" arrows would indeed seem to point towards INF economics: the full-employment and human-potential arrow, the healthy-environment arrow, the gradual-transition-away-from-fossil-fuels arrow and even the frugal-use-of-mineral-matter arrow.

Actually, the destination that these arrows really point towards goes by the generic name of "frugality".

Frugality

For the most part, we no longer even pay lip service to frugality. We have more or less dropped the word from our vocabulary, although as recently as 1978 Warren Johnson wrote a thoughtful book titled *Muddling Toward Frugality*. With one part of our brain, we have persuaded ourselves that frugality has no relevance to modern living. We no longer recite or even think about the old saying: "Waste not, want not." It sounds too old-fashioned and smacks of the Great Depression of the 1930s.

From one point of view, our collective decision to stop using the word "frugality" and to ignore the wastefulness built into conventional economics probably makes a certain amount of sense. After all, we are surrounded by enormous quantities of waste and garbage and scrap and junk and rust and so on, not to mention our excesses in matters of urban concrete and suburban asphalt and commercial ugliness. How could we possibly go out and do a normal day's work if our eyes were wide open to all these depressing sights around us? We probably need to keep our eyes and minds focused elsewhere just in order to retain our sanity.

What makes good sense for the short term, however, can be counter-productive in the long term. Willful blindness may help get us through the next couple of weeks. But ultimately, it only prolongs the agony and postpones the day of reckoning.

Unless I am totally mistaken, what happens when whole societies turn their backs on frugality and embrace modern wastefulness is that an insidious demoralization sets in. Little by little, we stop caring. We stop caring for future generations and we stop caring for the environment. We stop caring about children, the elderly, the weak and the unfortunate. Little by little, we lose the feeling of belonging to a community and we gain the feeling of living in a cold harsh world that shows no mercy to anyone.

Frugality is not really quantifiable. That is perhaps an additional reason why our number-loving society has so little use for it. But we all know what the word means. And we can all recognize frugality when we see it, just as we can all discern its almost total absence in the modern world.

As I see it, frugality does not preclude frivolity or spontaneity or enthusiasm or laughter or a sense of humour. It does not preclude art or music or drama. It does not even preclude a little bit of genuine extravagance from time to time, such as an ice cream cone on a hot day. But it does preclude the mindless ongoing promotion of consumption for consumption's sake. In a frugal society, one would not expect to hear the following kind of economic commentary: "Consumer confidence remains weak. Our exports, fortunately, are booming. But the economy cannot be said to have returned to good health until the domestic market regains its buoyancy. More consumer spending is essential."

Frugality asks, first and foremost, "Do people have what they need? If not, why not?"

Conventional economics asks, first and foremost, "Are consumers and businesses spending more money this year than last year? If not, why not?"

In order to understand how and why the modern world and frugality have parted company so completely, it is helpful to make a distinction between two different kinds of wastefulness. One kind is the familiar "consumer wastefulness" or "common wastefulness". It, in my view, is of secondary importance. The other kind, although of much greater importance, has no name. Let us call it "structural wastefulness".

Consumer wastefulness needs little discussion. We all know what it is and we all participate in it to some degree. What we do not always realize, however, is that the heart of modern wastefulness lies elsewhere.

Indeed, the heart of modern wastefulness lies in structural wastefulness. The latter can be defined as the wastefulness that results inevitably from a price structure where the monetary cost of being wasteful is often less than the monetary cost of frugality. Structural wastefulness follows from an unsound price structure in the same way that, as economics textbooks never tire of telling us, efficiencies follow from a sound price structure. Price structure is thus the key to structural wastefulness and, in fact, to modern wastefulness in general.

Examples of structural wastefulness abound in the modern world. A mechanic discards in good faith a malfunctioning component, not because he or she is incapable of repairing the component but rather because under conventional economics the monetary cost of such repair work would exceed the cost of a brand new replacement component. A huge advertising sign, illuminated all night long, provides us with no useful information but merely urges us to quench our thirst with a particular brand of beverage. A commuter finds it less expensive to drive an hour or more each way every day than to live close to his or her place of employment. Non-perishable as well as perishable goods are shipped overseas by air rather than by sea because all sorts of commercial advantages more than compensate for the somewhat higher freight rates. (See the Introduction and Chapter 5.) And so on.

Structural wastefulness is intimately linked to conventional economics. It is intimately linked to the distorted price structure that results when a society taxes human energy very heavily and non-renewable energy very lightly. And it is intimately linked to the fact that for competitive reasons no business in a profit-oriented economy can afford to be frugal where frugality carries a substantial financial penalty.

Not surprisingly, structural wastefulness seems to bring in its wake a certain amount of consumer wastefulness as well. We all find it difficult to resist the throwaway prices that are so common in certain parts of today's marketplace.

On the other hand, many retired people revert to frugality once their employment careers have ended. They repair what needs repairing, they re-use what deserves to be re-used, they design little household and backyard systems that are ingeniously frugal. No doubt one of the reasons for this reversion is that a retiree's income is generally independent of the amount of time that he or she spends on these kinds of frugality projects. But that is only an enabling reason. There must surely be a motivating reason as well, namely the fact that we all experience genuine pleasure and satisfaction when we indulge in a little creative frugality.

Back in 1899, Thorstein Veblen wrote the following somewhat convoluted sentence: "The popular reprobation of waste goes to say that in order to be at peace with himself the common man must be able to see in any and all human effort and human enjoyment an enhancement of life and well-being on the whole." Admittedly, the wastefulness that Veblen had in mind was the kind associated with conspicuous consumption. But perhaps his comment is equally applicable to the structural wastefulness that I have been describing. Perhaps there are even similarities between two sets of psychological circumstances: (a) those that lead to conspicuous consumption on the part of the wealthy and (b) those that push us all towards an acceptance of, and a willing participation in, modern energy extravagance. In short, perhaps Veblen is telling us, long after his death in 1929, why today we are not truly at peace with ourselves.

Deep down, I think, we all understand that frugality constitutes a major guidepost for the whole of human civilization. But we have allowed ourselves to be seduced into retreating farther and farther away from that guidepost. Two agents in particular (with the second being dependent upon the first) have practised this seduction with great success. First prize goes to conventional economics with its constant harping about our supposed need for (perpetual) economic growth. (See Chapter 13.) And second prize goes to the doubly subsidized advertising industry. (See Chapter 20).

Lastly, I would like to quote a sentence that not only contains the word "frugality" but also sums up much of what INF economics is all about. The sentence appears in the chapter focused on Herman Daly in the book *Wisdom for a Livable Planet* written by Carl N. McDaniel: "By setting limits on [the] production and consumption [of fossil-fuel and nuclear energy], we would have a frugality-first policy in which energy prices would go up, less energy would be used and for fewer things, and throughput would go down, while the market would efficiently allocate the energy allowed to it."

In my view, the promotion of sensible frugality constitutes a worthy criterion by which to judge economic systems and economic philosophies. Conventional economics, so judged, falls flat on its face, does it not? INF economics, I submit, at least remains standing erect.

15

Supply, Demand and Price

"In effect, the law of supply and demand works backward. Price becomes a primary economic fact, not a fact to be explained. By restating the law, making price the independent variable on which supply and demand depend, we have opened up consequences both powerful and liberating."

George P. Brockway (1993)

"The depletion of these critical [terrestrial] resources must therefore be rendered as small as feasible. Technological innovations will certainly have a role in this direction. But it is high time for us to stop emphasizing exclusively ... the increase of supply. Demand can also play a role, an even greater and more efficient one in the ultimate analysis."

Nicholas Geogescu-Roegen (1976)

The law of supply and demand

Everyone understands the essence of the law of supply and demand. Supply, demand and price all tend to have an effect on each other. If the supply of beef goes up, for example, then the price of beef will probably come down. And if the price comes down, then demand will probably go up. On the other hand, if a large new tax gets levied on beef but not on competing foods, then demand for beef will probably come down.

Despite being referred to as a law, the law of supply and demand does not really qualify as a scientific law at all. It is merely a convenient phrase that sums up some fairly obvious ways in which supply, demand and price tend to be interrelated. Depending upon the specific circumstances, those relationships may be quite tight, moderately tight or quite loose.

Two meanings of "demand"

It is easy to confuse two different meanings of the word "demand". Like many other words, the word "demand" started out having nothing but a plain ordinary meaning, and then subsequently it acquired a technical meaning as well. As often occurs in such cases, the difference between the two meanings is large enough to be important but small enough to permit confusion.

Here is an example of the word "demand" used with its ordinary meaning: "Her demand for a brand new replacement bicycle was unfairly rejected by the man who had driven his car over her old one."

Here is an example of the word "demand" used with its technical meaning: "Demand for new bicycles increased spectacularly after the price of gasoline doubled in less than six months."

We need to have names for these two different meanings.

Let us refer to the ordinary kind of demand as "fist-pounding demand", since it is often associated with feelings of indignation, with insistence on one's rights and with a literal or metaphorical pounding of one's fist on the table.

Let us refer to the technical kind of demand as "economic demand", since the word here is being used as a technical term in the context of economics. Where economic demand is involved, no one pounds on anything and no one insists on anything. Economic demand simply refers to the situation where, in the economic marketplace, there are people and/or institutions ready, willing and financially able to purchase the item in question at the price in question.

"Yes, yes, yes," I can hear the reader mutter impatiently, "but nobody would ever confuse those two meanings." I am not so sure. Consider crude oil, for example.

Crude oil

There is much talk nowadays about the supply of, and the demand for, crude oil. Supply, some people are warning us, may sooner or later not be able to keep up with demand. But which kind of demand are they referring to? To economic demand? Or to fist-pounding demand? Are they warning us that the various oil companies around the world are not going to be able to supply the world with as much crude oil as the world would dearly like to have and in fact is counting on having (fist-pounding demand)? Or are we being warned that the various oil companies around the world are not going to be able to supply the world with as much crude oil as the world is ready, willing and financially able to buy at the going price (economic demand)?

The answer here might actually be fist-pounding demand, rather than economic demand. After all, the oil companies can easily prevent or overcome any imbalance between supply and economic demand by the simple expedient of raising their asking price until economic demand no longer exceeds, or is no longer likely to exceed, supply. The people warning us are surely not worried that the oil companies will not be able to raise their selling price high enough to keep economic demand in balance with supply! No, they are worried that severe disruption and perhaps even chaos may lie ahead because the world may suddenly no longer have access to the quantities of crude oil that it has been anticipating and counting on. In other words, no matter how insistently we pound our fist on the table, no matter how indignant our feelings, no matter how genuine our recent trust of those experts who promised us a rosy future, we may not get what we desperately want.

As I see it, "demand" is not the right word to use in this context. Supply is not going to sink below demand, at least not below economic demand. That is not the problem. But on the assumption that the warning is valid, supply is going to sink below something else. It is going to sink below our wants. It may even sink below our short-term needs, since patterns of living, no matter how wasteful and extravagant, cannot always accommodate a major change overnight.

This phenomenon of supply suddenly sinking below our wants and perhaps even below our short-term needs will probably manifest itself (if it does occur) by a precipitous rise in the market price of crude oil. As a result, large numbers of people (manufacturers, consumers, hospital administrators, research scientists, snowplow owners, sports teams, journalists, truckers and countless others) may no longer be financially able to purchase as much crude-oil-derived energy as they had been counting on. At the same time, the oil companies and their shareholders may quietly rejoice at the windfall profits coming their way. In the language of supply and demand, supply will fall, price will increase, and demand will then fall as well so as to remain in balance with supply. Demand (i.e. economic demand) will not get ahead of supply, not if price is allowed to rise freely.

"Well, perhaps you are right," the reader may be conceding. "Perhaps some of us have indeed been confusing the two different kinds of demand. But as a practical matter, how important is all this? We are not interested in attending a class about word meanings. We are interested in energy problems and in possible solutions thereto. Stay focused, please!"

I will try to do so.

First, let me offer a general reason for paying attention to word meanings. The whole business of trying to think clearly and to communicate one's thoughts to other people becomes much more difficult whenever one finds oneself skating smoothly back and forth, without fully realizing what one is doing, between two different meanings for the same key word. (If the skating back and forth happens to be done cynically and with the deliberate intent to confuse and manipulate, then the behaviour becomes downright offensive.)

Consider, for example, the following two sentences that I have excerpted from an advertisement placed by a large oil company in the November 2013 issue of *Corporate Knights* magazine: "In the next 20 years the global demand for energy is expected to grow by more than one-third. So how do we find better ways to responsibly provide the energy we need so everyone can benefit from a healthy environment today and tomorrow?"

Which kind of demand does the company have in mind here? Is it fist-pounding demand? If so, just who is it who is doing the expecting and on what basis? Or is it economic demand? And if so, why is there no reference to price? After all, economic demand always refers to a particular price. Economic demand tends to go down as price goes up. So how can anyone predict future economic demand for energy without specifying anything about price? I have already pointed out how easy it would be for the oil industry to simply raise its prices in order to keep supply and demand in balance.

As I see it, the real key to this advertising message is to be found in the phrase "the energy we need" in the second of the above two sentences. The advertiser clearly wants us all to believe (a) that the world needs all the energy that we are currently consuming and (b) that the world will need 33% more energy in twenty years' time. The word "demand" serves as an artifice to distract us from the fact that global energy consumption is currently extravagant beyond words. Even if we consider the poorer countries of the world to be entitled to consume somewhat more energy per capita than at present, there is still plenty of room for a substantial decrease in global energy consumption, i.e. for a substantial decrease in both (a) annual global energy supply and (b) annual global economic demand for energy. No oil and gas company has any excuse for trying to make us believe otherwise. (But what are the chances that a committed believer in frugality, no matter how competent and hard-working, could ever survive long enough in today's corporate world to become the CEO of a large oil and gas company and change the company's basic philosophy with respect to growth in physical output?)

In discussions about possible future energy shortages, the meaning of the word "demand" is particularly important. Consider crude oil. Let us assume that the world's oil companies are currently extracting as much crude oil per day from the planet as they reasonably can. Let us further assume that global "demand" for crude oil is increasing significantly. What happens next?

What happens next, I submit, depends in large measure upon which meaning of the word "demand" we happen to have in mind. If economic demand has been placed in the spotlight, then a total of three different options will present themselves to us for consideration. But if fist-pounding demand holds sway, then one of those three options will probably get overlooked. My explanation follows.

Suppose that economic demand is being referred to here. In other words, buyers around the world are ready, willing and financially able to buy more and more crude oil as long as the price does not go up. According to what we all call the law of supply and demand, three basic options are available here in order to keep supply and economic demand in balance with each other.

One option is for all the oil companies to redouble their efforts to find crude oil and extract it from the Earth. Such a redoubling of effort might be motivated by financial hunger, by national or corporate pride, by governmental prodding or by some combination thereof. In the supposedly ideal case, global supply would successfully be increased, any price increases would be minor and would merely reflect any increased costs, and the increase in global economic demand would be matched by the increase in global supply. In a very general way, the world has mostly been following this first option for many decades. But recently, a second option has come on stage.

The second option is for all concerned to conclude that no satisfactory increase in the global annual supply of crude oil is realistically possible. In that case, the oil companies can simply raise the price of crude oil. Or it might be that most of their customers will feel obliged to bid higher prices than before. Either way, the price goes up. If the price increase is large enough, then global economic demand will fall, or at least stop rising, and supply and economic demand will once again be in balance.

One problem with this second option is that it can easily direct large windfall profits into the hands of oil companies and their shareholders. Another problem is that sudden large increases in the price of petroleum products may cause severe hardship for people with low incomes.

The third option, the one chosen by Intelligent National Frugality (INF) economics, is for national governments to examine the situation and to conclude that the world price for crude oil is unhealthily low and that, in consequence, economic demand in the wealthier countries of the world has become unhealthily high. But rather than allowing oil companies and their shareholders to receive undeserved windfall profits, each national government can impose a sizable tax on all crude oil entering the national economy. Such a tax would be designed to keep economic demand reasonably low and under control, as described throughout this book. Supply and economic demand would thus remain in balance with each other, albeit at a much lower level than before.

As long as the word "demand" is understood to mean economic demand, these three options remain on the table, and each one can be discussed on its merits, especially the third option.

In actual fact, however, consideration is seldom given to the third option, in spite of its very real attractions. The modern world seems almost exclusively focused on the first option. The modern world also seems quite unwilling to acknowledge that the first option, if pursued doggedly, cannot help but lead sooner or later to the second option. That is what finiteness and non-renewability are all about. As for the third option, well, it is simply beyond the pale!

Why is it that the third option receives so little consideration? No doubt there are many reasons (including the various explanations offered in the Introduction), but here I wish to suggest that confusion over word meanings may play a role. Specifically, it is easy to forget that the law of supply and demand deals with economic demand and not with fist-pounding demand. The difference is crucial.

In discussions about fist-pounding demand, we all tend to feel that in a free country anyone can demand—not necessarily obtain, but at least demand—whatever he or she wishes. That being the case, I have no business telling you what you should or should not be demanding. If you choose to demand a full barrel of crude oil (or the energy derived therefrom), who am I to say, "No, no! Half a barrel is all you are entitled to!"? In short, our right to demand is fundamental. It goes hand in hand with freedom of speech. Our demands may not always get satisfied, but our right to put them forward should not be restricted. If the third option goes against that right, then the third option deserves to be buried in an unmarked grave!

But neither the third option nor the law of supply and demand have anything to do with fist-pounding demand, at least not directly. The third option involves making use of the taxation system so as to raise the buyer's price for certain kinds of energy. That in turn will have an intended downward influence on economic demand. But no infringement of anyone's fist-pounding right is involved at all.

Admittedly, if the third option is adopted, it will probably cause society as a whole to alter not just its collective economic demand but also its collective fist-pounding demands. That too is intended. No one, for example, is likely to express any fist-pounding demand for the construction of additional lanes of highway in cases where automobile traffic is getting noticeably lighter with each passing year. On the other hand, we will all still be free to pound our fist on the table and demand that the potholes or ruts in such-and-such a roadway be filled in without delay. Our economy will change, but not our basic rights.

As for the question of how the third option can raise the buyer's price for petroleum products without causing undue hardship for people with low incomes, please see Chapter 4.

Electricity

In my view, our confusion over the intended meaning of the word "demand" is by no means limited to the topic of crude oil. It abounds in discussions of various economic topics, notably electricity demand.

Just as with crude oil, there is fist-pounding demand for electricity and there is economic demand for electricity. Fist-pounding demand for electricity is unusually strong for two somewhat related reasons.

The first reason is that, having grown up in the modern world, we all feel entitled to have lots of electricity at our beck and call. To use a thought-provoking term, electricity would seem to have become our favourite kind of "energy slave". Ivan Illich used this term in his 1974 book *Energy and Equity*. And Andrew Nikiforuk has thoroughly explored this slave analogy in his 2012 book *The Energy of Slaves: Oil and the New Servitude*. According to Nikiforuk, Buckminster Fuller coined the term "energy slaves" in the early 1940s.

When we turn on a light switch, even in broad daylight, we feel strongly entitled to have the light bulb receive its normal amount of electricity immediately. If that does not happen, we expect that someone will investigate promptly and will see to it that the problem is rectified without delay. This feeling of entitlement obviously could not have existed a few generations ago. But it does today.

The second reason is that, because of the complexity of the modern world, our lives would become almost impossibly difficult if we suddenly found ourselves deprived of electricity for an extended period of time. Indeed, for many of us, our basic need for food, clothing and shelter has come to include a basic need for a significant amount of electricity as well. For better or worse, that is a fact of life at present.

Given this fact of life, our society is obligated, for the time being, to provide everyone with enough electricity to keep body and soul together under today's peculiar social and economic conditions. One way to do this is to set such a low price for electricity that virtually everyone can afford to buy as much electricity as he or she needs. That is essentially the approach taken by conventional economics. But along comes the law of supply and demand to say, "Well, if you are going to set a very low price, then you are going to have to deal with a very large demand (i.e. economic demand), not just from consumers but also from the business world. In order to meet that very large demand, you will need to ensure that supplies are kept very large as well."

Indeed, conventional economics sets electricity prices so low that they strongly encourage the most extravagant, frivolous, wasteful, antisocial and environmentally harmful uses of this supremely convenient form of energy. One sees the tip of this particular iceberg, but only the tip, in the enormous electricity consumption of the advertising industry (television, the Internet, illuminated outdoor advertising signs of all kinds) and the entertainment industry (television, movies, special effects, casinos, theme parks, amusement parks, spectator sports palaces). And below the surface, super-cheap electricity is guilty of the worst kind of thievery; it steals countless good jobs away from human energy and allocates them instead to vastly less energy-efficient (as I will be discussing in Chapter 17) industrial uses of electricity in fields as varied as tailoring, furniture-making and boat-building. How does conventional economics justify this extraordinary state of affairs? It does so on the grounds that you and I can thereby afford to keep the food in our refrigerator cold!

As I hope I have shown, the law of supply and demand explains the basic economics of the modern world's absurd overconsumption of electricity. The same law also suggests a two-pronged alternative approach, such as the one taken by INF economics.

One prong would raise the buyer's price for most electricity to a level that is high enough to keep overall demand (i.e. economic demand) appropriately low. That in turn would enable supply to be kept appropriately low as well.

The other prong would involve some specific economic mechanism designed to ensure that even people not having much money could afford to buy enough electricity to live decent lives. Put another way, the second prong would respect one portion, and one portion only, of today's fist-pounding demand for electricity, namely the portion that corresponds to real physical need. For the INF version of the second prong, see Chapter 4.

Human energy

Nowadays, except in emergency situations ("Help! Police!" "Call the fire department!" "Get a doctor!"), few of us express much fist-pounding demand for human energy. Slave-owners and other such people may have done so in the past. But times have changed.

Economic demand for human energy, however, is alive and well. No, I exaggerate. It is alive but ill. Conventional economics does not seem able to establish a consistent economic demand for the right quality and the right quantity of human energy. It loves salespeople, for example, but only if they are exceptionally successful. It loves corporate deal-makers, but only if the corporate deals they put together are exceptionally large. It loves celebrities, but fame is inherently exceptional. It loves people willing to work for very low wages, but it generally rewards them with long hours, mindless work and dead-end jobs. It loves people able to make businesses grow, but it nearly always forgets to subtract job losses (e.g. those in a multitude of small downtown shops) from job gains (e.g. those in a handful of new "big box" stores and new suburban shopping malls); that is why one can say of the modern world, "So many new jobs have been created over the last several decades that there are now no longer enough to go around!"

Sadly, very sadly, conventional economics tends to shun those people, no matter how skilled and how conscientious, who would prefer to use mostly hand tools plus their own human energy to make beautiful and functional contributions to the economic and social health of their family, their community and their country. I have tried to express that defect more positively in Chapter 3.

As analyzed in this book, the basic problem under conventional economics is that the buyer's price for human energy is far too high when compared with the buyer's price for other forms of energy, notably fossil-fuel and nuclear energy. When price is far too high, economic demand generally falls to a level that is far too low. Nothing could be more straightforward.

The obvious way to rectify this problem is to get rid of all those taxes that have the effect of raising the buyer's price for human energy. Then as price comes down, economic demand can go back up again. Actually, economic demand for human energy gets a double boost when such taxes are eliminated. It gets boosted directly by the lowering of the buyer's price for human energy. It also gets boosted indirectly by the substantial raising of the buyer's price for fossil-fuel energy, since the latter often competes vigorously with human energy in the energy marketplace. The premise underlying this indirect boost is that some sort of new tax, such as the INF tax, is to be imposed on fossil-fuel energy as a replacement for the eliminated taxes on human energy. (For the other side of this coin, i.e. for the double penalty that conventional taxes inflict upon human energy, see Chapter 16.)

I picture the increased economic demand for human energy as occurring in thousands of different lines of work throughout the whole economy, not just in jobs requiring little or no skill. I also picture a widespread opportunity and incentive for everyone to improve his or her own personal level of skill and competence. And I picture more and more of us living and working more and more closely with nature: with farm crops and farm animals, with fish and marine life, with forests, with gardens, with wild animals and wild plants, with natural materials of all kinds and with solar and solar-derived energy. We will still need physics and chemistry and technical and engineering skills. But working in harmony with the subtleties of the natural world will become progressively more important as our reliance on mindless "energy slaves" diminishes.

Buyer's price and seller's price

As a moment's thought makes clear, "buyer's price" and "seller's price" are not always equal to each other. Under conventional economics, they are especially unequal as regards human energy.

Consider, for example, the typical case of Smith (a plumber) and Jones (an electrician) both of whom are trying to make a decent living under conventional economics. Suppose that the relevant income tax rate for both of them is 20%. Smith wishes to purchase $800 worth of electrical work from Jones and hence must go out and do $1,000 worth of plumbing work in order to obtain this after-tax amount of $800. But Jones only gets to keep $640 of this $800, since the other $160 goes to the government.

In other words, in order for poor Jones to end up with a measly $640, poor Smith has to go out and earn a much greater amount, namely $1,000 (which is more than 50% greater than $640). This perhaps explains why under conventional economics so many of us feel that we ourselves are underpaid when we sell our human energy and that those from whom we have to buy human energy are overpaid!

Indeed, one of the effects of a personal income tax is to open up a significant gap between the buyer's price for human energy and the seller's price. When Smith purchases human energy from Jones in the above example, the buyer's price (i.e. the price paid by Smith) is $800, while the seller's price (i.e. the price received and kept by Jones) is only $640. The difference goes to the government in the form of an income tax payment.

By eliminating personal income tax altogether, INF economics will close that price gap right back down to zero. If Smith pays $800 in order to buy some human energy from Jones, then Jones will be able to keep the full $800. The government will take nothing.

INF economics will actually alter these kinds of price gaps in two different ways. In the case of human energy, it will completely close the gap between the buyer's price and the sellers' price. But in the case of fossil-fuel energy resources, it will do the opposite; it will create a very sizable gap between the buyer's price and the seller's price, sizable enough to provide almost the totality of government revenue.

In the specific case of gasoline, we are already accustomed to a gap between the two prices, since small notices on gasoline pumps often tell us what proportion of the retail buyer's cost for a litre of gasoline goes to the government. INF economics will widen that gap considerably, not just for gasoline and diesel fuel but for all products derived from fossil-fuel energy resources.

Externalities

In economics, the word "externality" is used in situations where the price paid for a particular product or service or commodity does not cover all the real costs and consequences associated therewith. If I buy some gold bullion from a gold producer at an attractive price, for example, and if the producer later goes bankrupt and leaves behind a great deal of arsenic (As) pollution, then the cost of dealing with that pollution is referred to as an externality. Ideally, the pollution should have been prevented from occurring in the first place, and the cost of such prevention should have been "internalized" by being included in the price that I paid for my gold. But often, under conventional economics, that does not happen. The resulting externality, i.e. the resulting unfunded obligation to deal in one way or another with the gold producer's arsenic pollution, thus gets passed along to the government or to the public or to future generations. The gold producer and I share the benefit of the underpricing of the gold that I bought, while uninvolved innocent people get unfairly saddled with burdensome costs or hardships.

In any given situation, there are four basic approaches that a society can take in dealing with the problem of externalities:

1. One approach is to ignore the problem entirely and to let the chips fall where they may. That is what happened in the gold example above.
2. A second approach is to prohibit the whole economic activity in question. In the above example, society could simply prohibit all gold mining. In the case of nuclear power, society could formally ban the industry altogether, or it could use INF economics (or something similar) to achieve the same basic result.

3. A third approach is to pass and enforce strict laws and regulations ensuring that all proper production procedures are followed and that all relevant costs are internalized.

4. A fourth approach is for the government to do two things: (a) to impose an appropriately sized and appropriately calculated tax on the economic activity in question and (b) to use the ensuing revenue to nip any relevant externality problems in the bud. Clearly, this fourth approach involves opening up a gap between the buyer's price and the seller's price for the product or service or commodity in question. INF economics applies this fourth approach more or less automatically to almost every potential externality problem that it encounters.

It is worth noting that the third approach and the fourth approach are not mutually exclusive. A combination of both approaches might well make sense in many situations and would be readily accommodated by INF economics.

It is also worth noting that both the third and the fourth approach involve the law of supply and demand, since they both involve an increase in the buyer's price for the item in question. That increase might well cause overall economic demand to fall significantly, which in turn might significantly lessen the severity of the problem being dealt with. In certain cases, the increase in the buyer's price might even cause overall economic demand to fall all the way to zero. As mentioned, that seems virtually certain to happen in the case of nuclear power, once INF economics gets fully established.

Two externality examples

For a reason that I will explain in a moment, I would now like to discuss two important externality examples offered in 2003 by Vaclav Smil.

Smil's first example is that in the United States the current price of gasoline does not reflect the "cost of military and political stabilization of the Middle East". His thinking here is that, in a properly designed economic system, gasoline buyers should have to pay all the costs of producing the gasoline they buy, including an appropriate portion of the cost of "stabilizing" the Middle East countries where much of the crude oil (from which the gasoline is refined) originates. Otherwise, general taxpayers have to pay that particular cost, while the gasoline purchasers alone get the benefit. Not only do gasoline purchasers alone get the benefit, but also the low prices they pay have the effect of encouraging them to consume gasoline somewhat casually and thoughtlessly. In order to correct these defects, Smil and others argue (and I agree completely) that the Middle East cost should be internalized in one way or another.

Smil's second example relates to "the toll of respiratory illness caused and aggravated by photochemical smog created by car emissions". Here, the price of gasoline currently externalizes certain health costs that ought to be internalized because of their direct connection to gasoline consumption.

Smil's point in offering these two examples is somewhat different from, but by no means in contradiction with, the point that I am about to make. Smil's point is that in recent years a great deal of needed internalization of costs has already taken place in various sectors of the economy; but more still needs to be done, he feels, notably in the two examples he gives.

My point is somewhat different. My point is that INF economics, if adopted by the United States, would automatically internalize the externalized Middle East cost in Smil's first example. It would also automatically internalize a certain portion of the respiratory illness costs in Smil's second example (depending upon the extent to which the United States allocates health care to the public sector of the economy).

Moreover, if one combines the two automatic internalizations resulting from the application of INF economics to Smil's two examples, the total increase in the buyer's price for gasoline might be large enough to cause a major reduction in overall gasoline consumption and therewith a major reduction in smog-related illness. It might even be large enough to cause a major reduction in the perceived need to "stabilize" the Middle East.

My line of reasoning is as follows. Under INF economics, almost the totality of government revenue is to be obtained from the new tax on fossil-fuel energy resources. If extra government revenue is going to be needed in order to pay for certain government expenditures relating to the Middle East, then the INF tax will have to be increased accordingly. Under INF economics, there is simply no other acceptable source of revenue available. Similarly, if extra government revenue is going to be needed in order to pay, in whole or in part, for the prevention and treatment of smog-related illness, then here too that same INF tax will have to be increased accordingly.

Once these Middle East costs and illness-related costs have been internalized into fossil-fuel energy prices, those people who wish to reduce their own personal payment of such costs can effectively do so by reducing their own personal consumption (direct and indirect) of fossil-fuel energy resources. What could be fairer?

Remaining problems

Not even INF economics will automatically internalize all relevant costs and consequences. In particular, it will not automatically internalize those costs and consequences that our society might be tempted to leave for future generations to deal with. Moreover, INF economics has no way of doing the impossible, namely of internalizing the cost of those far-in-the-future liabilities associated with the disposal of long-lived nuclear wastes; the only solution for that problem is (a) to stop producing such wastes and (b) to try to deal as wisely as possible with those wastes that we have already produced.

Nevertheless, as I see it, INF economics still has two aces up its sleeve. With one ace, it will so encourage frugality and energy efficiency, that very few of these future-generation problems will likely even arise. With the other ace, it will help create a less acquisitive and less power-hungry society in which the very idea of thoughtlessly and greedily short-changing future generations will seem morally repugnant.

Summary

In economic matters, it is important to do the following six things:
1. To avoid any confusion between economic demand and fist-pounding demand.
2. To remember that price can be used in many cases to move economic demand up, if desired, or down, if desired.
3. To remember that supply and demand can be kept in balance with each other not only by a deliberate and

carefully planned increase in the one but also by a deliberate and carefully planned decrease in the other.

4. To keep in mind that, in any given situation, buyer's price and seller's price are not necessarily equal to each other.
5. To ask oneself whether, in any given situation, there are any externalities that need internalizing.
6. To keep in mind the basic immorality involved when one generation of human beings knowingly, selfishly and uncaringly creates serious problems, or serious potential problems, for future generations.

16

The Price of Fossil Fuels

". . . there is no such thing as the cost of irreplaceable resources . . ."

Nicholas Georgescu-Roegen (1972)

"It is apparent that resource prices are to a large extent arbitrary—a fact that is seldom recognized."

Herman E. Daly (1991)

This chapter puts forward two basic arguments. The first is that there is no such thing as a "true" price for fossil-fuel energy resources. And the second is that, in the absence of a "true" price, it makes sense to choose a wise price.

The absence of a "true" price

What makes fossil-fuel energy resources absolutely unique is the non-renewability of the chemical energy they contain. Once the chemical energy stored in a litre of gasoline or in a cubic foot of natural gas has been converted into heat and then used for some intended purpose or purposes (or simply wasted), the energy is gone for good and can never be recaptured. Admittedly, the carbon (C) atoms and hydrogen (H) atoms that make up the molecules found in such fuels do not disappear. But for all practical purposes the energy does, because the energy ultimately ends up as useless waste heat. Hence both the original fossil-fuel resource itself and the chemical energy it contains are said to be non-renewable.

233

(Solar energy and the various kinds of solar-derived energy are also destined to end up sooner or later as useless waste heat. But because fresh solar energy is continuously arriving on our planet, everyone accepts the legitimacy of categorizing solar and solar-derived energy as renewable energy.)

Because of the non-renewability of fossil-fuel energy resources, the total existing quantity of such resources on the planet Earth keeps getting smaller and smaller as we destroy them—barrel by barrel, cubic foot by cubic foot, trainload by trainload—for the purpose of exploiting the chemical energy they contain.

Put another way, the oil companies do not "produce" any oil at all. If they did, then the total existing quantity would not keep getting smaller and smaller. Such companies provide a service, the service of extracting and refining oil and transporting it and so on. But they do not "produce" it. No one does. It is not getting "produced" at all. It is simply getting extracted, as I mentioned in Chapter 14.

What then are the "true" (i.e. market-determined) prices for oil and coal and natural gas? My answer, and that of many other people, is that no such prices exist. The market is simply incapable of determining, all by itself, the price of non-renewable commodities. This inability stems from logic, not from practical difficulty.

Now obviously one can go out and buy some gasoline on the market and be quoted a precise price for the transaction. All over the world, prices for fossil-fuel energy resources do in fact exist. But my point is that in no case have those prices been determined solely by the market. Markets simply cannot do that, either in theory or in practice.

Why can they not? Part of the answer is that, in order for markets to be able to play their assigned role in the setting of prices, both supply and demand must be able to function in the normal manner. No one, however, is capable of bringing to the market any new supply of fossil-fuel energy resources.

Some readers, I am sure, are shouting out at me, "I disagree! New discoveries of oil are constantly being made. New supplies of oil are constantly being brought to market." My response is that new discoveries of oil do not constitute new oil. Any and all newly discovered oil is essentially just as ancient as the oil discovered and brought to market fifty or a hundred years ago.

New discoveries of oil constitute new knowledge, not new oil. The same is true of new technology that improves our ability to discover and exploit oil deposits. Knowledge can of course be very valuable. But knowledge and oil should never be confused with each other. In particular, the fact that our geological and technological knowledge is constantly increasing should never be allowed to obscure the fact that the total quantity of raw petroleum oil on this planet is constantly decreasing.

Conceptually speaking, all fossil-fuel energy resources belong in the same category as a painting by Vincent Van Gogh. They are unique, one of a kind, irreplaceable and therefore priceless.

What about the auction process? Clearly, the market is capable of putting a price on a Van Gogh painting by means of a public auction. There is a crucial difference, however, between the auctioning of a valuable painting and the auctioning of a batch of coal or oil or natural gas.

In the case of a valuable painting, everyone agrees that the ultimate goal focuses on preservation: preservation of the painting itself and preservation of its value. But in the case of a non-renewable energy resource, the ultimate goal is to bring about the destruction of the resource, usually via combustion. Given such a fundamental difference, everyone—including members of future generations—who might like to enjoy the benefits of having, say, some natural gas to burn ought to be able to bid on the item in question. Otherwise, the auction is not open to all concerned and hence cannot be said to be fair, and any prices set by the auction cannot be said to have been truly and properly set by the market.

Since there is no way in which future generations can place their bids, the unavoidable conclusion is that no auction in the world is capable of setting a truly market-determined price for a fossil-fuel energy resource. Moreover, any auction that does happen is likely to place an artificially low price on the resource, precisely because so many potential bidders (members of future generations) have necessarily been excluded from the bidding process.

I am not saying that the auction process can never be used as part of a mechanism for setting prices for fossil-fuel energy resources. I am simply saying that, even if the auction process does get used for such resources, the resulting prices can never be described as being set by the market alone.

In making this same argument against the apparent legitimacy of allowing auctions by themselves to set prices for non-renewable resources, Nicholas Georgescu-Roegen has written as follows: "There is an elementary principle of economics according to which the only way to attribute a relevant price to an irreproducible object ... is to have everyone bid on it ... [But] [f]uture generations are not, simply because they cannot be, present on today's market."

As it happens, there are many different ways in which prices for fossil-fuel energy resources can be set. None of them, however, involve exclusive determination by the market and hence all of them are best described as being arbitrary.

One option, for example, would be for the national government to create a national state-owned corporation having a domestic monopoly over the extraction and sale of raw fossil-fuel energy resources. That corporation could then set prices for all such resources at almost any level it wished, subject only to whatever guidance the national government saw fit to provide. The corporation—or the government itself—might even decide to leave a particular fossil-fuel deposit undisturbed in the ground for, say, 100 or 500 years, so as to give some future generation the opportunity to exploit the resource.

Another option, the one advocated in this book, would be for the government to impose an Intelligent National Frugality (INF) energy tax. Expressed in dollars per unit of energy, that tax would constitute the basic price of all raw domestic fossil-fuel energy (except for the rationed amounts of tax-free fossil-fuel energy discussed in Chapter 4). The wholesale and retail prices for refined fossil fuels such as gasoline and natural gas would then include (a) the basic price as set by the INF tax, (b) additional amounts covering all the various services provided (extraction, storage, transportation, refining, distribution, financing, marketing and so on) and (c) profit for the entities providing those services (assuming that profit is allowed).

Many other scenarios can also be imagined.

Does all this theorizing about market inability really matter? I think it does because it means that none of us can hide our own energy-pricing preferences behind the supposedly objective operation of the marketplace. Instead, each of us has to take a stand. Each of us has to say, in effect, "I realize that ultimately the price of fossil-fuel energy resources is arbitrary. Our society must therefore decide which arbitrary price to choose. Here is my own personal recommendation, together with my underlying reasons."

As a society, are we not actually very fortunate in being able to make a choice for ourselves on such an important matter? Market inability here is surely deserving of a smile rather than a frown!

A wise price for fossil-fuel energy

Once we are in agreement that there is no such thing as a "true" or market-determined price for fossil-fuel energy, we are left with the problem of deciding what would constitute a reasonable or wise price for such energy.

Under INF economics, the government openly acknowledges the arbitrary nature of the basic price that it sets for fossil-fuel energy. That basic price takes the form of a fossil-fuel energy tax expressed in dollars per unit of energy. Being so expressed, the tax is completely unconnected to any of the other components that make up the final retail price of, say, a litre or gallon of gasoline. As for the tax rate itself, it gets arbitrarily set at a level that is just high enough to yield almost all the revenue needed or wanted by the government.

Now, is that price too high, or is it too low, or is it just right? Is it a foolish price or is it a wise price?

It is best, I think, to approach this question conceptually. If we use numbers, they will drown us.

We can perhaps all agree that energy constitutes the key ingredient in all economic production. If energy is unavailable, nothing gets produced, not even a wild blueberry. For the latter to grow, solar energy is needed.

In the case of virtually every conceivable commodity and product and service, there are choices to be made concerning which kind or kinds of energy to use and how much of each kind to use. Even in the case of wild blueberries, one might or might not decide to make use of a certain amount of both human energy and fossil-fuel energy (the latter converted into the chemical energy of "gunpowder") by firing gunshots towards any hungry black bears observed in the vicinity of the berries.

Enter economics. In the modern world, once a human being makes the decision to try to play a useful role in some productive activity, his or her energy choices will for the most part be determined by price and availability. In particular, those choices will be determined to a very large degree by the relationship between the cost of human energy and the cost of fossil-fuel energy. (For the sake of simplicity, I am mostly ignoring nuclear energy in this chapter.) Society therefore needs to ask itself, "What is the optimum relationship between the price of human energy and the price of fossil-fuel energy?"

Here we want an answer in the following form: "The optimum relationship between these two energy prices is the one most likely to cause such-and-such to happen." And we need to fill in the "such-and-such" blank.

I suggest that "such-and-such" has two components: a human component and an environmental component. We therefore need to consider our question from those two different points of view.

Let us discuss the human point of view first. According to one of the most fundamental teachings of our culture, each human being is unique and has a unique potential. In order for us to live up to our potential, we each need to have the opportunity to lead a productive life. Our overall potential is by no means confined to matters of work and production. But neither is it

239

totally divorced therefrom. It tends in fact to get stunted if we are deprived of the opportunity to perform meaningful work under conditions of reasonable dignity and fairness.

From the human point of view, therefore, the optimum relationship between the two energy prices in question is the one that maximizes the opportunity for each of us to perform meaningful work under conditions of reasonable dignity and fairness. We do not mind using our bodies and our muscles up to a point, but we do not want to work as beasts of burden all day long. Nor as mindless robots. (Back in the 1970s, I once watched a factory worker doing "mindless robot" work tending a stamping press in an automotive plant. It was a depressing sight and doubtless humiliating for the Charlie Chaplinized machine-tender.) Nor do we want to be relegated to the economic sidelines while others roll up their sleeves and pitch in. We want to be satisfyingly productive.

We therefore need an economy that discriminates, in the best sense of the term, in favour of human beings and human energy. As I stated in Chapter 1, we need an economy that turns to all of us as human beings and says to us, "You humans are entitled to choose your jobs first. I will not offer any work to any other forms of energy until you have made your choices. Naturally, the more highly skilled among you (or in some cases those who are willing to put up with more hardship, risk or discomfort) will get the better-paying jobs. But there is plenty of work available for everyone, precisely because I am granting to you human beings the right of first refusal with respect to nearly all jobs in the whole economy. Take your pick. Furthermore, I encourage you to continue developing your skills and talents throughout your working life and to take on new projects and new responsibilities from time to time if you feel so inclined."

In order for all this to happen, the present gap between the high cost of human energy and the low cost of fossil-fuel energy will have to be narrowed considerably. Otherwise, we human beings will have to continue contenting ourselves with the absurd "right of second refusal" that we now possess with respect to job offerings. For is it not true that our present economy tends to offer as many jobs to nuclear and fossil-fuel energy as is technologically possible, regardless of how many human beings might be willing and qualified to accept those jobs?

When was the last time, for example, that you heard of someone in a modern industrialized country being paid to cut grass with a non-motorized lawn mower? Under INF economics, by contrast, lawns in city parks can be expected to get mowed by a phalanx of people of various ages, all pushing expertly designed, expertly made and expertly sharpened pollution-free mowers powered exclusively by human energy. The same kind of mowers will no doubt also get used on the lawns of private homeowners, pushed either by members of the household or by paid help. As for rural lawns, some will surely get grazed, in part at least, by sheep or other domestic animals.

And why, nowadays, does so much of our new furniture get manufactured in large, centralized, automated, energy-guzzling factories and then get transported at further energy cost over large distances, instead of being hand-crafted locally by the human energy of skilled cabinetmakers and upholsterers? Clearly, the answer centres around price, not around quality, creativity, originality, craftsmanship, resource conservation or energy efficiency.

So we need to figure out, still conceptually speaking, the amount by which the price gap between human energy and fossil-fuel energy ought to be narrowed. Not eliminated, just narrowed. (I will elaborate on this point in a moment.)

A useful conceptual technique for dealing with problems of this kind is to start at the bottom of the ladder and work up, or else start at the top of the ladder and work down. For convenience, let us start at the top of the ladder and work down. For the ladder itself, let us choose the price ladder for fossil-fuel energy. So the top of the ladder represents high prices for fossil-fuel energy, while the bottom represents low prices.

At the topmost rung of this ladder, the price of fossil-fuel energy would be so prohibitively high that the annual total consumption of such energy would be zero. Such a price would certainly succeed in discriminating in favour of human energy. It would also result in a major gift of non-renewable resources to future generations. But Intelligent National Frugality is focused not on rejecting fossil-fuel energy entirely but rather on making use of such energy in a wise and frugal manner. So the topmost rung of the ladder is not for us.

The second rung down from the top of the ladder would be the one where the price of fossil-fuel energy is set high enough, by means of a fossil-fuel energy tax or something comparable, so that absolutely all government revenue gets provided by the tax in question. At first glance, that might appear perfect. It certainly would have the effect of discriminating in favour of human energy. But it would leave no room for the government to generate revenue by, for example, imposing fines as punishment for crimes and other offences.

As I mentioned in Chapter 2, not all government revenue has as its sole purpose the financing of government expenditures. Criminal fines are just one example. An inheritance tax would be another, if one of its purposes was to keep a lid on overall wealth inequality. So would an income tax payable only by the very wealthy. In all such cases, the special purpose involved might well be considered sufficiently important to override the basic goal of discriminating in favour of human energy. Moreover, the total amount of special revenue involved in these cases would probably be small enough to have little or no overall effect of an undesirable nature on energy prices and energy choices.

Since the second rung down is still not entirely satisfactory, let us descend to the third rung. At this rung, the price of fossil-fuel energy is set high enough, again by means of a tax on fossil-fuel energy or something comparable, so that all government revenue, except the special revenue discussed in the preceding paragraph, gets provided by the tax in question. In my view, this third rung is the perfect place to stop. At this third rung, we can most effectively discriminate in favour of human energy, without in any way compromising the ability of the government to carry out its functions. (For the sake of simplicity, I am overlooking in this chapter the points I made in Chapter 6 about municipal governments.)

Note that neither at this third rung nor even at the second rung does the price of fossil-fuel energy have to have caught up to and surpassed the price of human energy. That will inevitably happen someday, as exploitable fossil-fuel energy resources become increasingly scarce. (So eventually, but not for a very long time if we act promptly, future generations will still get short-changed by being deprived of ready access to such resources.)

But for the time being, in the advanced industrialized countries of the world, human energy will remain more costly than fossil-fuel energy under virtually all scenarios. What I am discussing in this book is not "Which of these two kinds of energy ought to be the more expensive?" but rather "Should the price gap between these two kinds of energy be narrowed significantly?" Accordingly, to discriminate in favour of human energy means to narrow the price gap from its present width. It does not mean to raise the price of fossil-fuel energy above that of human energy.

That said, we need to confirm that the third rung is indeed the right place to stop. So let us continue the exercise by descending even further down the ladder. We should expect to find that all lower rungs pull us step by step farther away from our desired goal.

Suppose, then, that we move a few more rungs down the ladder. It does not really matter which particular rung we select. Suppose we stop at the rung where the price of fossil-fuel energy is only set high enough (by means of the new tax) to bring in one-half of the ordinary revenue required by the government. In that case, the other half would have to be generated by some other tax or taxes: perhaps a sales tax and/or a personal income tax and/or a corporate income tax.

Now the key characteristic of taxes such as those just mentioned is that they do not discriminate between human energy and fossil-fuel energy, at least not by sparing the former. In their (blind) eyes, production is production, sales are sales and a dollar is a dollar; the kinds of energy consumed are irrelevant. In consequence, such taxes will inevitably tax human energy some of the time. And when they do, human energy gets doubly penalized. It gets penalized by having its own price raised. And it also gets penalized by having the price of its principal competitor (fossil-fuel energy) lowered.

Let me be very clear about how this double penalty gets inflicted upon human energy and about how the price of its principal competitor gets lowered. Obviously, a sales tax (for example) does not, by itself, lower the price of anything. It certainly does not, by itself, lower the price of fossil-fuel energy. But we are not talking here about an abstract "by itself" situation. We are talking about the real world where governments need to obtain their revenue from one kind of tax or another. The choice is not between, say, a sales tax and no tax. Rather, it is between a sales tax and a fossil-fuel energy tax. And the key point is that the effect of the sales tax when compared with the effect of the fossil-fuel energy tax is to raise the price of human energy and to lower the price of fossil-fuel energy.

Let us return to our ladder of fossil-fuel energy prices. The problem with all the rungs below the optimum one can now be seen clearly. None of them discriminate as strongly in favour of human energy as does the optimum rung. In other words, they all have a tendency to take good jobs away from human beings and to turn those jobs over to fossil-fuel energy. As far as I can see, there is nothing particularly beneficial about any of these lower rungs. They all penalize human beings. The lower the rung, the greater the penalty.

Let us now leave the human point of view and shift to the environmental point of view, while still considering the question of the optimum relationship between the cost of human energy and the cost of fossil-fuel energy. As it happens, both points of view yield the same answer. They do so because in both cases it is desirable to discriminate in favour of human energy and against fossil-fuel energy. Indeed, the environmental impact of productively used human energy is generally quite minor when compared with that of our fossil-fuel competitors. (Please keep in mind that, as I have already pointed out, human beings have to be fed, clothed and sheltered regardless of whether their human energy is put to productive use or not.)

So environmentally speaking as well, the optimum rung on our ladder is the third one from the top: the one where the price of fossil-fuel energy gets set high enough, by means of a fossil-fuel energy tax or something comparable, so that all government

245

revenue except the special revenue mentioned earlier gets provided by the fossil-fuel energy tax.

This new high price for fossil-fuel energy will create a very strong incentive for everyone—producers, consumers and governments—to do three things: (a) to turn first to human energy, solar energy, solar-derived non-human energy (such as wind energy and the chemical energy in wood and biofuels), tidal energy and geothermal energy before turning to fossil-fuel energy, (b) to use fossil-fuel energy with a very high degree of efficiency and (c) to use fossil-fuel energy not just with great efficiency but also with frugality. The environmental benefits that will flow from such a strong incentive are obvious.

Just in case any reader is still unconvinced, let us descend a few rungs further down the ladder, as we did earlier. In doing so, we would weaken the threefold incentive outlined above. We would thus start working against the environment in three different ways:

1. In some situations, we would no longer turn first to human energy and solar energy and so on. Instead, we would go straight to fossil-fuel energy. For grass-cutting, for example, we might go straight to gasoline-powered mowers without even considering any non-motorized alternative. And for most of our furniture-making, we might go straight to centralized energy-intensive assembly-line production, bypassing our pool of local cabinetmaking and upholstering talent.

2. Our incentive to obtain a very high level of energy efficiency from each and every use of fossil-fuel energy would be weakened. Adding extra insulation to our houses in order to minimize wintertime heat loss, for example, might simply not be cost-efficient and might therefore not get done.

246

3. Our incentive to use fossil-fuel energy frugally would also weaken. As householders, for example, we might pay less attention to keeping our home and our workplace reasonably close to each other. And as manufacturers, we might succumb to the temptation to consume electricity generated from fossil-fuel energy for the purpose of illuminating a few extra billboards advertising our products.

For all the above reasons, then, I submit that INF economics will impose on fossil-fuel energy a price that is necessarily arbitrary but at the same time wholly consistent with our most important social and human values. If that claim is justified, then surely this new price deserves to be called a wise price.

17

Efficiency

"Our century has been slow to grasp the truth that the efficiency of an action has absolutely nothing to do with its rightness or wrongness."

David Ehrenfeld (1993)

"A key task in addressing today's economically framed dilemma is to expose and redress a basic error of industrialism: pursuing narrowly defined mass-production efficiency by substituting capital, energy, and resources for human inputs of skills and labor."

Hazel Henderson (1996)

Preamble

Although there are various kinds of efficiency, they all have to do with some sort of a comparison between output and input. With high efficiency, the ratio of output (or other desired outcome) to input is high. With inefficiency, the ratio is low. Irrigation water for farm crops, for example, is said to be used efficiently in circumstances where almost all of the water contributes directly to the healthy growth of the crops in question. But if most of the water simply evaporates or runs off without helping plant growth at all, then that particular irrigation system is said to use water very inefficiently.

This chapter deals mostly with two different kinds of efficiency. Regardless, however, of the kind of efficiency that one has in mind, efficiency should never come first. Efficiency should always play second fiddle to kindness, empathy, decent behavior and high ethical standards. Otherwise, one risks ending up with the terrible efficiency of such abominations as slavery, the Nazi death camps and the deliberate genocidal distribution of disease-ridden blankets, as well as countless lesser evils.

Two kinds of efficiency

Two different kinds of efficiency get talked or written about frequently. One kind is energy efficiency. The other is cost efficiency. Much of the time, however, we neglect to specify, and sometimes even to understand, which kind we are referring to. Endless confusion results.

The crux of the problem is that these two kinds of efficiency do not always go hand in hand. They do sometimes, but not always. Often, especially under conventional economics, option A will be more cost-efficient than option B, while option B will be more energy-efficient than option A. If the kind of efficiency being discussed is left unspecified, misunderstandings are almost unavoidable.

Cost efficiency

Cost efficiency refers to the costs and benefits—all measured, wherever possible, in dollars or some other currency—that will be generated if a particular course of action is taken. If several different options are being considered, then the one that produces the greatest economic benefit per dollar spent, or the one that accomplishes the desired result at the lowest financial cost, will constitute the most cost-efficient option. An insurance agency, for example, may have to decide whether to computerize all its customer files rather than hiring additional office personnel.

This being a business matter, the decision will almost certainly be based on cost alone. Whichever option gets the required job done for the least amount of money will get the nod.

Cost efficiency generally sits at the head of the business world's table, as long as the actions being taken are neither illegal nor ethically unacceptable. Energy consumption, apart from its monetary cost, does not enter into calculations of cost efficiency.

In one important respect, cost efficiency gets determined in a manner that is ultimately arbitrary. Once one accepts the argument that the basic price of fossil-fuel energy is inherently arbitrary, as discussed in Chapter 16, it becomes apparent that virtually all prices in a modern national economy are to a certain extent arbitrary. Cost efficiency is therefore itself based to a certain extent on the arbitrary assumptions and arbitrary legislated decisions of the national economy in question. Put another way, what is cost-efficient under one approach to economics may not be cost-efficient under a different approach. In particular, what is cost-efficient under conventional economics will often not be cost-efficient under Intelligent National Frugality (INF) economics. And vice versa.

More than by any other profession, cost efficiency gets examined and evaluated by the accounting profession. (Honourable mention can perhaps go to those who manage to make ends meet despite having to, or choosing to, live on a very small cash income!)

Energy efficiency

Energy efficiency is very different. Its measurement does not depend upon money and prices at all. And energy efficiency is not the least bit arbitrary. Quantities of energy count, whereas quantities of money do not. The energy efficiency of a particular electric motor, for example, can quite easily be calculated. If during the course of one hour the motor receives 100 units of electrical energy and puts out 90 units of mechanical energy, the energy efficiency of the motor is 90%. Similarly, if a coal-fired generating station puts out 30 units of electrical energy for every 100 units of chemical energy contained in the coal being burned, the energy efficiency of this conversion of chemical energy into electrical energy is only 30%.

By the laws of science, energy efficiency as defined in this manner cannot exceed 100%. In fact, it cannot even attain 100%. If one wishes to transform the chemical energy in gasoline into the kinetic energy of a horizontally moving automobile, for example, there will always be some energy losses in the form of waste heat, especially from the hot engine itself. The energy efficiency of any internal-combustion engine is always substantially less than 100%. And the same is true for every other kind of heat engine.

More than by any other profession, energy efficiency gets examined and evaluated by the engineering profession. (Honourable mention can perhaps go to those for whom horsepower normally means *horse* power!)

Energy efficiency and task accomplishment

Another way to look at energy efficiency is to consider a particular task and try to identify the most energy-efficient method of accomplishing that task. Consider, for example, the task of transmitting large quantities of electricity over long distances. At what voltage can such transmission take place with the highest degree of energy efficiency? Electrical engineers know that transmission losses are much smaller under conditions of high transmission voltage than under conditions of low transmission voltage. They therefore recommend that transmission voltages be raised as high as possible, although not so high that other problems appear.

By definition, the most energy-efficient way of carrying out a particular task means the way involving the least consumption, or the least loss, of energy. If less energy is lost when long-distance transmission lines carry 250,000 volts than when they carry only 100,000 volts, then clearly for this particular task the former voltage offers greater energy efficiency than does the latter.

One of the most attractive ways to increase energy efficiency is to use the same energy to accomplish two or more separate tasks simultaneously. Car-pooling—using the same energy to transport two or more unrelated people to their destination—constitutes an obvious example. So does co-generation, which refers to the combustion of natural gas (or some other fuel) in special power plants that are designed to generate both electricity and useful heat simultaneously.

An easily overlooked way of increasing overall energy efficiency is to eliminate the need for, or the desirability of, accomplishing certain tasks in the first place. None of the various methods of illuminating an advertising billboard at night, for example, can even come close to the energy efficiency of never illuminating the sign at all. Similarly, no air conditioner or commercial motor vehicle or high-rise office building or sightseeing helicopter can match the energy efficiency of the one that never came into existence in the first place. And no new kilowatt or megawatt of electrical generating capacity can get used with as high a degree of energy efficiency as the "negawatt" of capacity that never gets built or installed in the first place.

Strictly speaking, I admit, it makes no sense to speak of the energy efficiency of something that does not exist or is not happening. In such cases, both the numerator and the denominator of the output-over-input fraction are equal to zero; and zero divided by zero does not produce a meaningful number. Nevertheless, we all need to be aware that our society's overall energy efficiency can be significantly and intelligently improved if we simply stop engaging in certain energy-consuming activities where the benefit-to-cost ratio, expressed not in dollars but in real social and environmental benefits and costs, is unreasonably low. For example, a great many of the modern world's air-freight shipments, as I have suggested in the Introduction and Chapter 5, constitute transportation absurdities that have a very unsatisfactory benefit-to-cost ratio, once one looks beyond the dollars involved.

This leads me to smallness. Consider, for example, the Alberta oil sands. The exploitation of the oil sands requires, among other things, massive processing facilities called "upgraders" that upgrade the raw bitumen into synthetic crude oil. The latter can then get fed into normal oil refineries for conversion into gasoline, diesel fuel and so on. Clearly, the amount of upgrading capacity needed, whether located nearby or far away, depends upon the average daily quantity of bitumen being extracted from underground. Far more upgrading capacity is needed to handle, say, a million barrels of extracted bitumen per day than to handle one hundred thousand barrels per day. (For the sake of simplicity, I will continue with that ten-to-one ratio in the following paragraphs.)

One might argue that, once they are built, the energy efficiency of ten identical upgraders is the same as for a single one of such upgraders. True enough. But that narrow argument ignores the fact that far more energy is needed to construct ten identical upgraders than to construct just one. The only reason to construct and operate the other nine is that under conventional economics the commercial world is in a terrific hurry to consume crude oil (synthetic or otherwise) as fast as it possibly can. Unbelievably, as I have already mentioned, the human race consumes approximately a billion barrels of crude oil every couple of weeks, with the wealthiest countries of the world, including Canada, taking far more than their fair share on a per capita basis.

Admittedly, the supply of exploitable Alberta bitumen might outlast the useful operating life of a single upgrader, even one designed to be readily repairable and to last as long as possible. On the other hand, what are the odds that under conventional economics all ten upgraders will be maintained and repaired and operated farsightedly right up until they are beyond repair? Do we not often hear about corporations closing—for economic reasons—factories and refineries and so on that are still in satisfactory operating condition?

In a very real economic sense, as I hope this oil sands example illustrates, haste makes waste. In our haste to exploit our bitumen resources as quickly as possible, we bear prime responsibility for wasting all the energy and other resources needed to construct nine-tenths (in this example) of the relevant upgrader capacity. And unless and until we bring our haste under control, the number of unnecessary upgraders is likely to keep on increasing, with a concomitant decrease in the overall energy efficiency of our oil sands exploitation.

My argument here is not confined to bitumen upgraders. It also applies to traditional oil refineries. And to fossil-fuel pipelines, as regards their combined total length, their average diameter and their associated machinery. And to a significant portion of the rest of our transportation infrastructure.

So it turns out that small is not just beautiful, as E. F. Schumacher argued persuasively in *Small is Beautiful: Economics as if People Mattered*. In a great many contexts, small is also energy-efficient.

For my comments on energy efficiency and task accomplishment with regard to the actual extraction of fossil-fuel energy resources from the Earth, see the section on EROEI below.

Maximizing energy efficiency (not quite)

An interesting, important and perhaps unexpected observation can be made at this point. Not even Intelligent National Frugality (INF) economics will seek to maximize the overall efficiency with which our society uses energy. Do not misunderstand me. INF economics will seek a very high level of overall energy efficiency, far higher than is the case under conventional economics. But it will not seek the very highest possible level. That would require a complete cessation of the use of all non-renewable energy resources, including all fossil fuels. INF economics simply does not go that far.

Let me explain. If a society truly wished to achieve a maximization of its overall energy efficiency, then that society would have to carry out the totality of its tasks with the least possible overall consumption of energy. But the totality of its tasks can be summed up as the one general task of operating a decent society for all its members. What, then, is the least possible overall energy consumption that is consistent with a decent society for everyone?

For our purposes here, we can reword that question as follows: What is the least possible amount of nuclear energy consumption and fossil-fuel energy consumption that is consistent with a decent society for everyone? Clearly, it is impossible to prove that such an amount is greater than zero. No one can come up with some number like 50 and claim that the human race could manage all right with 50 units of non-renewable energy but never with 49. After all, there is always room for improvement. Moreover, can anyone really believe that once all the exploitable nuclear and fossil-fuel resources on the planet are exhausted, the human race will be permanently condemned to a life that is nasty, brutish and short?

I stress this point so that none of us get ourselves into hot water by misunderstanding or misrepresenting what INF economics will do. It will not maximize our overall energy efficiency. Not quite. One cannot have one's cake and eat it too. One cannot consume even a small amount of fossil-fuel energy and at the same time claim that no further reductions in one's overall energy consumption are realistically feasible.

Conventional economics

How and why does conventional economics perform so poorly in matters of energy efficiency? It does so by financially discouraging a great many sensible economic activities that, if they were carried out, would be characterized by a high degree of energy efficiency. Put another way, conventional economics drives a huge wedge, a far larger wedge than in the case of INF economics, between energy efficiency and cost efficiency. It thereby sets up, time and time again, situations where ultimately no one can win. Those who opt for energy efficiency end up losing money. And those who opt for cost efficiency end up aggravating the many serious problems (human problems, resource-depletion problems and environmental problems) associated with excessive and inappropriate consumption of energy.

Admittedly, there are a number of situations where energy efficiency and cost efficiency do go hand in hand, even under conventional economics. Generally speaking, for example, it is now more economical, i.e. more cost-efficient, to install double-glazed windows rather than single-glazed windows for lessening wintertime heat loss. Similarly, it is often both cost-efficient and energy-efficient to use electronic controls to switch on and switch off furnaces, engine-block heaters, outdoor lights and so on at the desired time, rather than leaving them switched on all night long. And manufacturers of such complex items as jet aircraft engines and data-bank server computers are starting to incorporate improved energy efficiency into the designs of their products.

The successes of conventional economics, however, should not be allowed to hide its failures. These include all the transportation absurdities that I have mentioned previously. Moreover, there is one broad area where conventional economics performs abysmally with regard to energy efficiency. This broad area embraces all those situations in which conventional economics steals desirable jobs away from human energy and allocates the work to non-renewable energy instead. Quite apart from the resulting human distress, this blind preference for non-renewable energy engenders some almost incredibly inefficient uses of energy. I will discuss some numbers in a moment.

The essence of the matter can be stated quite simply. Conventional economics does pay a certain amount of attention to energy efficiency, but only after casting aside, wherever it can, one particular kind of energy, namely human energy. This bias against human energy has a drastic negative effect on overall energy efficiency.

Now for some numbers

First, let us obtain an approximate feel for the size of the gap—as created by conventional economics in the modern industrialized countries of the world—between the cost of fossil-

258

fuel energy and the cost of human energy. Let us do so by comparing two distances: (a) the distance that ten dollars' worth of gasoline will move a car along a level stretch of highway and (b) the distance that ten dollars' worth of minimum-wage human energy will move the same car. And note that the latter distance would become even shorter if the minimum-wage human energy were replaced by human energy costing, say, three or five or ten or fifty times more than the minimum wage.

My point here is not to suggest that it would ever make sense to use human energy for the purpose of propelling a car. (It often does make sense, of course, to use human energy for the purpose of propelling a bicycle.) Rather, my point is that under conventional economics the cost of fossil-fuel energy amounts to only a pittance when compared with the cost of human energy. Under conventional economics, human energy costs a fortune.

Without being overly concerned about numerical accuracy, let us say for the sake of argument that under conventional economics the cost of human energy in a modern industrialized country such as Canada is one hundred times higher than the cost of fossil-fuel energy, when the two prices are expressed in dollars per unit of energy.

That 100-to-1 ratio has an extremely negative implication for energy efficiency. Any employer or potential employer, including any government employer, can save money by discarding, say, one unit of human energy and replacing it with up to ninety-nine units of fossil-fuel energy. In other words, it can be cost-efficient to choose a level of energy efficiency that is only 1% as high as would be the case if human energy had been chosen. After all, ninety-nine units of fossil-fuel energy would still cost less than one unit of human energy! (For the sake of simplicity, I am ignoring here the fact that fossil-fuel energy, unlike human

259

energy, is not of much use unless some kind of machinery or other equipment gets deployed as well.)

As an example, think of all the complicated machinery in an automatic car-washing facility. And think of all the non-renewable energy that gets consumed in the manufacturing, the transporting and especially the powering of all that machinery. There is no comparison between the ridiculously poor energy efficiency of such a facility, measured in terms of total energy consumed per car-wash, and the far higher level of energy efficiency in a competing operation powered almost exclusively by human energy. But in business, cost efficiency takes precedence over energy efficiency. Under conventional economics, automatic car-washing is both cost-efficient and rapid and has therefore been able to take over much of the market from its vastly more energy-efficient human competitors.

Intelligent National Frugality (INF) economics

How will INF economics change things? In modern industrialized countries such as Canada, will it reduce the gap between the cost of human energy and the cost of fossil-fuel energy all the way to zero? No it will not, at least not for a very long time if our luck holds. Imagine how tiny a quantity of gasoline (refined from fossil fuels) each of us could afford if such gasoline became so expensive that ten dollars' worth did not move an automobile any further than ten dollars' worth of human energy! Sooner or later, fossil fuels being non-renewable, such a day will inevitably arrive. But one goal of INF economics is to postpone that day for as long as reasonably possible.

So without immediately reducing the above-mentioned gap all the way to zero, INF economics will narrow the gap considerably. It will do so both by lowering the cost of human energy (through the abolition of all taxes thereon) and by raising the cost of fossil-fuel energy (through the imposition of a major new tax thereon and through the abolition of all relevant subsidies.)

A narrowed gap will automatically reduce the incentive to consume energy inefficiently. Let us assume, again for the sake of argument, that under INF economics, the new ratio of the cost of human energy to the cost of fossil-fuel energy will start out at only 20 to 1 instead of 100 to 1. Accordingly, no employer or potential employer will have any incentive to replace one unit of human energy with any more than nineteen units of fossil-fuel energy. Moreover, employers will have an incentive to switch back to human energy in all those cases where, previously, they had substituted between twenty-one and ninety-nine units of fossil-fuel energy for a single unit of human energy.

By the same token, abundant opportunities for small-scale family farming will spring up like new trees and new forests after a lengthy ice age. Based on that same 20-to-1 ratio, family farms will be able to recapture food-production work from agribusiness in all those situations where agribusiness has needed more than 20,000 kilocalories of non-renewable energy, consumed directly or indirectly, in order to match the agricultural production of 1,000 kilocalories of human energy. (Kilocalories are not usually used to measure quantities of energy contained in fuels such as gasoline and diesel fuel. But, as I stated in Chapter 1, simple arithmetic can easily convert kilocalorie measurements into BTU measurements and so on.) Given the enormous energy appetites of agribusiness, fending off the INF challenge from family farms will almost certainly prove financially hopeless. Under INF economics, in short, agribusiness will not only be much less energy-efficient than the family farm, which is already the case, but also much less cost-efficient.

Regardless of whether one finds these numbers of mine helpful, it is important to grasp the basic principle involved. In a nutshell, the greater the ratio of the cost of human energy to the cost of fossil-fuel energy, the higher will be the level of energy inefficiency that remains cost-efficient. As far as I know, that basic principle is not mentioned in any standard economics textbook. Conventional economics simply does not allow itself to think about such matters.

Let me repeat myself for a moment. INF economics does not seek the absolute maximum in energy efficiency. It therefore does not seek to completely eliminate all fossil-fuel energy consumption. It does not even seek to completely eliminate the price gap between the high cost of human energy and the low cost of fossil-fuel energy. But what it does seek is a much-reduced gap between those two prices.

Once that reduction has been accomplished, fossil-fuel energy will no longer be in the driver's seat. Most of the starkly inefficient uses with which such energy is currently associated will cease, being no longer sustained by cost efficiency and cheap affordability. In particular, fossil-fuel energy will no longer be able to steal good jobs away from human energy. Fossil-fuel energy will no longer be able to promote, and have its own consumption promoted by, our seemingly insatiable desire for "more, bigger, faster". It will no longer be able to sell itself for a song and thus get itself used for any number of frivolous, absurd and anti-social purposes. And it will be increasingly unable to compete with the solar or solar-derived energy associated with hydroelectricity (modest-scaled), solar-generated electricity, wind power, a clothesline, an equator-facing window, a draft horse, firewood, other modestly produced biomass fuels and so on.

Moreover, from the point of view of society at large, the attractions of direct and indirect fossil-fuel energy consumption will no longer be quite so able as they are at present to overwhelm the attractions of walking, chatting, reading, canoeing, bicycling, horseback riding, skating on natural ice, snowshoeing, natural food, natural music, natural fabric, natural wood, live theatre (frugally produced) and so on.

Energy return on energy invested (EROEI)

Let me return for a moment to task accomplishment. One particular task, namely the task of extracting fossil-fuel energy resources from the Earth, merits special comment. And for simplicity here, I will single out the oil industry.

As with all endeavours in the business world, the oil industry tries to find and extract oil in as cost-efficient a manner as possible. That is why the industry has been picking all the low-hanging fruit first. But now that most such fruit has already been picked, the industry is allocating more and more of its investments to the task of extracting so-called unconventional oil, such as shale oil and Arctic oil and the bitumen in oil sands.

With unconventional oil, the industry still tries to maximize cost efficiency. But precisely because most of the lower-hanging fruit has already been picked, the cost of extracting an additional barrel of oil tends to keep increasing. By "cost", I mean not just the monetary cost but also the energy cost. In recent years, therefore, attention has begun being paid to an arithmetical ratio known as "energy return on energy invested" or "EROEI" (or sometimes shortened to "EROI").

Back in the days when many oil wells were described as gushers, almost no one cared about the EROEI ratio. It was clearly so high that the exact number was not worth calculating. Everyone knew that successful well-drilling yielded a huge return of energy. Profit margins, not EROEI ratios, monopolized everyone's attention.

Oil being a non-renewable resource, however, circumstances have gradually changed. In order to find and extract each additional barrel of oil, the oil industry is now having to invest more and more energy. This energy comes in two forms. One form consists of the energy embedded (or embodied) in ships, trucks, drilling rigs, well pipes and so on. The other form consists of the chemical energy contained in all the gasoline and diesel fuel and natural gas employed and consumed in the oil patch.

In short, EROEI ratios are getting lower and lower. But as a practical matter, they cannot keep on decreasing indefinitely. Once the ratio falls to one-to-one, no one will have any incentive whatsoever to continue extracting oil. The petroleum era will have passed into history.

Will INF economics have any impact on this situation? Yes, it almost certainly will. It will almost certainly raise the minimum EROEI ratio considered by the oil industry to be acceptable. Here is why. Oil exploration and oil extraction, like most business activities, necessarily involve a certain amount of gambling (rational gambling for the most part, one hopes!) Oil companies gamble that in general their investments in the oil patch will pay off. Under INF economics, however, the amount of money at stake will become larger and larger as the EROEI ratio falls lower and lower. Admittedly, that is also true under conventional economics, but not nearly to the same extent. Under INF economics, the progressive enlargement of the stakes involved will be vastly greater.

Imagine, for example, the situation where an oil company drills a dry (and therefore worthless) well in extremely inhospitable terrain such as the high Arctic. The well being dry, the investment will have to be written off. The gamble will have failed. But under INF economics, the amount of money unsuccessfully invested will have been enormous because the company will have previously paid a great deal of INF tax on two different categories of invested fossil-fuel energy: (a) the embedded energy mentioned above and (b) the chemical energy mentioned above.

Clearly, the very same oil-patch investment, measured in physical terms rather than in dollars, will cost much more under INF economics than under conventional economics, especially in those increasingly common situations where the amount of energy being invested is substantial. That basic fact will surely lead the oil industry to aim at a significantly higher minimum EROEI ratio under INF economics than under conventional economics.

Given the likelihood that EROEI ratios will trend downwards with each passing year, should INF economics be modified so as to offer some sort of tax rebate in connection with energy invested in the extraction of unconventional fossil-fuel energy resources? Not in my view. Our generation's task is to focus on intelligent frugality and to lower our consumption of fossil-fuel energy resources as quickly and substantially as reasonably possible. We do not need to encourage the energy industry to squeeze every last exploitable barrel of oil out of the ground.

Efficiency and human energy

Let me restate some points from Chapter 1. Human energy derives from the sun via the food we eat. And solar energy comes to us each day as a gift. So from the standpoint of energy efficiency, not to mention the Golden Rule (stated in Chapter 8), it makes no sense to organize our society in such a way that a certain amount of human energy remains involuntarily idle. That is the equivalent of saying to the sun, "You are supplying us with more energy than we can use. You are very kind and we appreciate your thoughtfulness, but we simply cannot figure out what to do with this daily energy surplus. Regretfully, therefore, we allow it to go to waste."

One can take the point still further. In a civilized country, we try not to let anyone go unfed or unclothed or unsheltered. Even those who, for whatever reason, produce little or nothing of value get provided with the necessities of life (nearly always). But food, clothing and shelter all represent energy expenditures on the part of society. Some of the energy expended might be solar-derived and some of it might not be. All of it, however, gets expended for the purpose of keeping all of us decently alive.

Here is a perfect opportunity for our society to increase its overall energy efficiency by using the same energy expenditure to accomplish two different tasks simultaneously. One task focuses on keeping the person in question decently alive. The second task can take the form of any worthwhile activity in which that particular person has significant competence. So in place of any involuntary unemployment or involuntary underemployment that might otherwise occur, society could offer itself extra medical personnel, extra playground supervisors, extra life guards, extra food inspectors, extra coast guard personnel, extra safety inspectors, extra police officers on foot patrol, extra crew members on railway trains, extra teachers, extra librarians and so on.

In most cases, as I have already mentioned, even people with serious handicaps have something of value to contribute to society, if given the opportunity.

Very little extra consumption of fossil-fuel energy need accompany this extra utilization of human energy. Under conventional economics, admittedly, the amount of fossil-fuel energy consumed by an individual doubtless tends to go up significantly as his or her income rises. Under INF economics, however, people with comfortable incomes will have a strong financial incentive to avoid allocating very much of their surplus cash to purchases, direct or indirect, of fossil-fuel energy. My guess is that such people will instead be much more likely to put money aside for their retirement years and/or spend their money on activities and purchases involving only small inputs of fossil-fuel energy. Accordingly, even if both a decent job and a reasonable wage get offered to those who might otherwise be involuntarily idle, our society's overall energy consumption will probably not thereby increase significantly.

An additional point relating to efficiency and human energy involves certain public buildings, notably public libraries. Under conventional economics, we have generally found it easy to keep our public libraries comfortably warm all winter long because of the low price of non-renewable energy. But in many cases, especially in small towns and villages, we are now finding that we cannot afford to keep the libraries open to the public for as many hours per week as we might like. The cost of the needed librarians' human energy is simply too great. In other words, much of the non-renewable energy that we consume in our libraries gets wasted because of the high cost of human energy! INF economics will make it much easier for us to employ sufficient staff to keep our public libraries and other public buildings open to the public for a generous number of hours per week.

Finally, nothing in this section or in this whole book should be taken to imply that the only legitimate alternative to involuntary idleness is paid employment. As every vegetable-gardener and every do-it-yourself enthusiast knows, one can be quite productive without earning any money. Moreover, one can engage in many worthwhile human activities, notably the raising of a family, without producing anything having market value at all. Money-based economic activity is important in the modern world, but it is by no means the only important thing in life.

Efficiency versus risk avoidance

Sometimes, especially under conventional economics, too much emphasis on efficiency promotes the taking of certain risks, risks that some people might regard as foolish. This brings me back to the need to assign no more than second-fiddle status to efficiency. Let me give two examples.

My first example concerns cost efficiency and long-distance food transportation. Under conventional economics, it often costs less money to buy certain kinds of food from far away, perhaps even from a different continent, rather than from local sources. But risks are involved when one's local community or one's local region comes to rely on distant sources for a good portion of its food supply. If unforeseen circumstances were suddenly to interfere, qualitatively or quantitatively, with that long-distance food supply chain, would there be enough time to establish adequate alternatives before people began suffering seriously? The answer to that question cannot be provided by economic theory. Economics cannot tell us which risks are foolish and which are acceptable. We have to make such decisions for ourselves, either as individuals or as a whole society.

If, as a whole society, we decide, as I personally feel we should, to proceed cautiously regarding food staples, then we will want to adopt appropriate policies for the promotion of a large degree of local food self-sufficiency in every part of the country. It so happens that INF economics by itself will go a long way towards accomplishing that goal. It will do so (a) by favouring the family farm and small-scale agriculture, because of their inherent energy efficiency, and (b) by imposing major financial costs on large-scale centralized agribusiness and long-distance food transportation, because of their inherent energy inefficiency. Once the cost efficiency of purchasing foods from far away has disappeared, the temptation to take the risks involved will also have disappeared. But let me state once again that we do not have to rely exclusively on economics in order to shape our society or determine which risks are worth taking. Like all independent countries, we in Canada have plenty of legislative and regulatory tools at our disposal as well.

My second example comes from Jane Jacobs. In her 2004 book *Dark Age Ahead*, she warns that excessive efforts to maximize the efficiency (she does not specify any particular kind of efficiency) of a whole country could easily lead to the slow strangulation of the country's basic culture. Culture, she feels, cannot get passed along from one generation to the next without a great deal of human mentoring and human nurturing. She refers to all this mentoring and nurturing activity as "redundancy". For my part, I would describe it as an abundant and diversified input of human energy directed, consciously or otherwise, towards cultural renewal and cultural promotion.

As far as I can see, there is no conflict at all between Jacobs' particular concern here and INF economics. The latter, by leaving human energy untaxed and therefore readily affordable, can be expected to offer a great deal of encouragement to all the "extravagant", "redundant" and "inefficient" mentoring and nurturing activities that Jacobs values so highly.

Here are two of Jacobs' key sentences on this point: "Perhaps the greatest folly possible for a culture is to try to pass itself on by using principles of efficiency. When a culture is rich enough and inherently complex enough to afford redundancy of nurturers, but eliminates them as an extravagance or loses their cultural services through heedlessness of what is being lost, the consequence is self-inflicted cultural genocide."

18

Productivity

"It is ironic that we define productivity as the elimination of labor in the manufacturing process, when it is human activity that can bring life back to our fields, forests, watersheds, and even our factories."

Paul Hawken (1993)

"One may want to contrast this notion of productivity—churning something out at the lowest cost whether anyone needs that something or not—with the notion of 'copeability,' the ability to deal and cope adequately with a variety of circumstances. Copeability is a quality much valued and respected in the women's world."

Ursula M. Franklin (1985)

One word, two meanings

The word "productivity" has a very positive ring to it. How could anyone have doubts about something whose desirability seems so obvious? But "productivity" does not always mean what one might think. In this chapter, I would like to discuss two of its meanings in particular. One of those meanings can be summed up by the words "personal skill and competence". The other meaning is "labour productivity".

271

Personal skill and competence

Based on our own observations of how individuals go about their daily work, we are likely to say, for example, that McDonald is a more productive plumber than McCall. In such cases, we use the words "productive" and "productivity" to refer to the personal abilities of individual human beings. Since we know that both McDonald and McCall have access to the same selection of tools, equipment, electricity and so on, we correctly attribute the difference in their respective outputs to a difference in their respective personal abilities and/or work habits.

This kind of productivity is clearly worth encouraging. Indeed, McDonald's high level of skill and competence as a plumber will normally benefit society as a whole as well as benefiting McDonald personally. That is a key reason why society willingly spends so much money on the education and training of children and young adults.

There are also perfectly legitimate feelings of pride and satisfaction that accompany the knowledge that one is highly competent at one's job. That is true even if the work in question is unpaid. In every household there is unpaid work to be done, and the skill and competence brought to bear on such work can make a huge difference to the quality of life of every member of the household.

In referring to personal skill and competence, we sometimes use the words "efficient" and "efficiency" rather than the words "productive" and "productivity". As long as the reference is unmistakably to personal ability and personal performance, however, the meaning is the same.

Where problems do arise is in the use of the word "productivity" in economics.

Productivity in economics

In discussions about economics, the word "productivity" is usually used to mean "labour productivity", although it can also refer to the productivity of some other input such as electricity or land or water or retail floor space or capital.

A summary definition of the term "labour productivity", as used in economics, is "output, measured in dollars (or other financial currency), per hour of human labour".

At first glance, that definition might seem reasonably close to the concept of personal skill and competence. After all, the average hourly output of a person possessing a high level of skill and competence will normally, other things being equal, be considerably higher in dollar value than that of a less competent person. But as it turns out, there is actually a world of difference between (a) the everyday concept of personal skill and competence and (b) the economic concept of "labour productivity". The difference centres around the different assumptions that are normally made in the two cases.

In the case of productivity based upon personal skill and competence, we normally assume that all other factors are the same for all the individuals whose productivity is being compared. Sometimes we even explicitly include the phrase "other things being equal" (as in the previous paragraph) so as to ensure that no misunderstandings are possible. Our principal assumption is that all the individuals concerned have equal access to the same selection of tools and equipment and non-human energy. Obviously, McDonald's productivity as a plumber would fall far below that of McCall if someone were to confiscate all of McDonald's tools and equipment. But precisely because that is so obvious, it is usually left unsaid.

In the case of "labour productivity" as the term is used in economics, the situation is generally quite different. Instead of focusing on personal skill and competence and instead of assuming that other factors are the same for everyone, modern economics tends to do the opposite. It tends to focus on the other factors and to take personal skill and competence more or less for granted. So when modern economics says that a truck-driver is more productive than a wheelbarrow-pusher, no one is talking about the skill and competence of either person. The truck-driver's productivity is high because of the truck, not because of any extraordinary skill on the part of the driver. And we need to remember that the truck itself is totally useless in the absence of a suitable sort of non-human energy such as the chemical energy stored in diesel fuel.

Once one shifts to the term "labour productivity", one has to be extremely careful about the conclusions one draws. For the unwary, danger lurks.

The principal danger lies in assuming that high levels of "labour productivity" are always a good thing, regardless of how they are obtained. That assumption is unwarranted. Each instance of high "labour productivity" must stand or fall on its own merits. One instance cannot be permitted—just because of the pleasant sound of the word "productivity"—to ride on the coattails of another instance. More specifically, high "labour productivity" cannot be permitted to bask undeservedly in the praise that we quite properly bestow upon high levels of personal skill and competence. If high "labour productivity" wishes to receive our praise, then in each case it must earn that praise.

The key point here is that "labour productivity" statistics, because of the very definition of the term, take no account of any of the non-labour inputs involved nor of any of the social and environmental consequences involved. In other words, such statistics take no account of a good portion of reality. But if one confines one's consideration to only a small portion of reality, how can one possibly form an intelligent judgment about what is going on?

Regardless, therefore, of whether "labour productivity" levels are rising or falling, we still need to ask pertinent questions. Is a particular economic activity environmentally benign? Are scarce resources being husbanded? Are high safety standards being met? Are employment levels satisfactory both quantitatively and qualitatively? Is the younger generation being well educated and well trained? None of these considerations are automatically promoted by rising levels of "labour productivity". Nor are they automatically compromised when "labour productivity" falls.

Let us now look at five examples.

Example one: two carpenters

The first example, which I have borrowed from the American economist George P. Brockway, shows that under certain circumstances everyone can receive an economic benefit—not just a social benefit but an actual economic benefit—even though average "labour productivity" actually falls.

Here is Brockway's scenario. A master carpenter, having been working alone up until now, decides to hire a semi-skilled assistant from the ranks of the unemployed. Because the skills of the new assistant are far from being equal to those of the master carpenter, the total hourly output (measured in dollars) of the two persons working together does not attain a doubling of the master carpenter's previous hourly output. In other words, average "labour productivity" goes down. But as long as the new assistant receives fair remuneration, everybody benefits. The master carpenter, the new assistant, government finances and society as a whole all benefit. (This assumes that the carpentry work in question is intrinsically worthwhile.)

It could be argued that Brockway's example represents a special case. After all, the paid-work productivity of the new assistant was previously zero (actually, zero divided by zero), since he or she was previously unemployed. It then attained a respectable level. "Labour productivity" statistics, however, follow the economic definition of "labour productivity" and hence do not take the unemployed into account. Accordingly, such statistics tell us that in Brockway's example "labour productivity" falls. The whole truth is thus quite different from the statistical half-truth.

Brockway himself sums up his point as follows: "As my grandmother used to say, every little bit added to what you've got makes a little bit more. The nation does not become stronger or richer by keeping any potential worker unemployed. The notion of productivity is macroeconomically irrelevant."

Brockway's point can be taken even further. If we simply wanted to raise the level of our country's "labour productivity", one way of doing so would be (a) to keep the country's overall output at a constant level (measured in dollars) and (b) to keep laying off additional workers (i.e. human energy) and replacing them with a combination of capital equipment and non-renewable energy. Such a deliberate promotion of unemployment would of course be highly antisocial. But the economic statistics would indicate that "labour productivity" was constantly increasing. The ultimate goal would be to have only one person supervising a huge army of robots doing all the work in the whole country!

Jerry Mander comments on these matters as follows: "Robotics, computers, and biotechnology are three major reasons that workers are becoming redundant. The substitution of those machines for human beings is conventionally called 'gains in productivity.' And it does increase productivity in one sense, but it throws a lot of people out on the street."

Example two: city buses

For a second example, consider the case of city buses. At one time, it was normal for city buses to be operated by a two-person crew: a driver and a conductor. The driver saw to the operation of the bus as a motor vehicle, while the conductor dealt with the passengers and with all matters relating to fare payment. Then, in many cities, the decision was made to dispense with bus conductors. Two-person crews were replaced with one-person crews. So in addition to their previous responsibilities, drivers were assigned all the responsibilities (apart from making change) that had formerly been carried out by conductors. As a result, the average "labour productivity" for city bus crews instantly doubled.

But the fact that "labour productivity" doubled does not tell the whole story at all. What happened to the former conductors? Did they simply join the ranks of the unemployed (or push others into those ranks)? How much additional stress was imposed upon the now-unassisted drivers? Was the safety of the bus passengers compromised? Was the safety of nearby pedestrians and cyclists compromised?

In cases like these, safety and "labour productivity" are often perched at opposite ends of a teeter-totter. When one goes up, the other goes down. Edward S. Cassedy and Peter Z. Grossman have made the same point in connection with coal mining: "But better mine safety and health provisions not only have a cost in terms of new equipment, they also have a cost in worker productivity."

No doubt there are limits to the number of personnel that can reasonably be hired for the purpose of enhancing safety. But we are not likely to get very close to those limits if we insist that falling or non-rising "labour productivity" levels constitute by themselves a sign of poor economic performance.

Example three: executive travel

For my third example, let us return to the situation (mentioned in Chapter 1) where business executives are required to travel long distances as part of their corporate duties. Many business executives at present earn very large salaries and bonuses. In order to get their money's worth for all this financial compensation, the companies in question need to maximize the "productivity"—output per hour of human-energy input—of their executives. One's "productivity" tends to be low, however, when one is in the process of travelling, even if one has access to computers and other such devices. So a minimization of executive travel time constitutes an obvious corporate policy to pursue.

Two approaches suggest themselves here. One approach, the one normally taken under conventional economics, is to minimize executive travel time by maximizing executive travel speed. Hence the use of corporate jets and the supersonic Concorde (until it was withdrawn from service in 2003). By contrast, the normal approach under Intelligent National Frugality (INF) economics will be to minimize executive travel time by minimizing executive travel distance. This will occur almost automatically, given the decentralized, frugal and generally smaller-scale nature of the INF economy.

The difference between these two approaches is fundamental. The high-speed approach that is encouraged by conventional economics creates a number of problems. Scarce fossil-fuel energy resources get squandered, large amounts of carbon dioxide (CO_2) get emitted into the atmosphere, and various other kinds of environmental degradation occur as well. The resulting "productivity" of the executive may be high, but so are the environmental price and the resource-depletion price that society as a whole has to pay.

Example four: blister packs

We are all accustomed nowadays to purchasing small hardware items that have been pre-packaged in blister packs. The latter generally consist of a clear plastic "blister" glued to a boxboard backing. From the point of view of both trash reduction and resource conservation, nothing complimentary about such blister packs can be said at all. Why then does conventional economics find them so attractive? The answer is that in two different ways they raise the "labour productivity" of retail salespeople.

First, by transforming small easily concealable merchandise items into much larger and more conspicuous ones, blister packs tend to discourage shoplifting. Secondly, they often reduce the need for labour-intensive counting or weighing or measuring. Both factors help the retail store in question to get by with fewer salespeople. And by the logic of "labour productivity" calculations, output per hour of salesperson labour must go up if total output remains constant while the total number of hours of salesperson labour goes down. Such calculations ignore all the social and environmental costs associated with blister packs. So, as often happens, "labour productivity" goes up, while society as a whole loses out.

Example five: product design

My fifth example is wholly contained in the following quotation from Paul Hawken (with emphasis in the original): "By designing products so that they can be disassembled and remanufactured, we will require more labor, a cost that will ultimately be paid for by using less waste and energy. This is one example of how, in the restorative economy, productivity can go *down*, employment up, and profits increase."

Summary

Under conventional economics in particular, "labour productivity" statistics do not even begin to tell the whole story of how well a particular economy is performing. Furthermore, no necessary correlation exists between high "labour productivity" and desirable economic activity. The former is no more likely to be associated with the latter than with the exact opposite of the latter.

Perpetual growth

Generally speaking, conventional economics believes in perpetual "labour productivity" growth, just as it believes in perpetual economic growth. Both beliefs are purely ideological. Both sound foolish when the word "perpetual" is explicitly stated. And both lie at the heart of the problems being discussed in this book.

Actually, the two beliefs boil down to one and the same belief. They do so because perpetual economic growth is supposed to keep ahead of population growth and because perpetual economic growth is not supposed to require that each year we all work longer and longer hours. So the kind of economic growth usually envisaged is the kind that occurs when we all keep on working steadily and, with much help from our supplies of non-renewable energy while they last, we all produce a little more per hour this year than we did last year. In short, we want our desired perpetual economic growth to result from perpetual growth in "labour productivity".

Given the unity of the above two ideological beliefs, all the criticisms that this book directs against the pursuit of perpetual economic growth can also be directed against the pursuit of perpetual growth in "labour productivity".

Productivity and residential housing

While uninterested in "labour productivity" statistics, INF economics places a great deal of importance upon the other kind of human productivity, the kind attributable to personal skill and competence. Such productivity goes hand in hand with the creative and productive best in human beings. In order for such productivity to be promoted properly, three things are necessary. First, high levels of personal skill and competence must be attained by each new generation of the population. Secondly, those high levels of skill and competence must then be fully applied to the tasks at hand. And thirdly, the tasks in question must be sensible and worthwhile.

In my view, conventional economics often fails the second and third requirements. It does so by causing human energy to be overpriced and therefore, in many situations, inadequately or inappropriately applied.

Consider, for example, the construction of residential housing. Under conventional economics, a team of architects, backhoe-operators, foundation-builders, carpenters, roofers,

281

bricklayers, plumbers, electricians and so on may design and build houses capable of lasting, say, seventy-five to one hundred years. Not very long at all, given the amount of resources involved. Indeed, writing in 1968, the British economist Charles Carter somewhat wistfully made the following observation: "In 1900, any substantial dwelling was expected to have a life of indefinite length, perhaps of many centuries. Now [i.e. nowadays] it is sometimes considered a virtue or advantage that a building should have a built-in obsolescence which will require its replacement after sixty or eighty years."

Something has gone wrong here. What, if anything, has happened to the personal skill and competence of our house-building labour force? And can it really be true that the "labour productivity" of that same labour force has somehow been going up, even while the longevity of our stock of residential housing has been going down? My comments follow.

Regarding the first of those two questions, I doubt that anything at all has happened to the personal skill and competence of our house-building labour force. Those people do what they are paid to do, and they work within the time frames expected of them by their employers. Most of them are probably quite capable of building houses that would fully meet Charles Carter's longevity standards. But no one is asking them, or paying them, to do so. Any prospective homeowner leaning in that direction would have to be prepared to pay as many as three extra bills: (a) the architect's bill for a custom-designed long-life house, (b) the bill for the extra labour involved and (c) the bill for the extra materials involved. Under conventional economics, the first two bills in particular might be quite large.

Regarding the second question, it appears that the "labour productivity" of house-building tradespeople has indeed been going up, even while the longevity of the houses they build has been going down. How strange! For an explanation, let us return to our basic definition: output (in dollars) divided by the number of hours of labour input. Based on that definition, "labour productivity" must rise if labour input falls while output does not fall. Put another way, "labour productivity" must rise if the same result can now be accomplished with fewer hours of labour than before. And clearly, during the past half-century or so, labour input for a typical new house in countries such as Canada has in fact been decreasing.

It is the same old story. Labour input has been decreasing because labour (i.e. human energy) has been receiving more and more assistance from, and undergoing more and more displacement by, the combination of non-renewable energy and mechanized tools and equipment: power tools, earth-moving machinery, hydraulic lifting devices, concrete delivery trucks, roof-truss manufacturing plants, roof-truss delivery trucks and so on. Moreover, being very expensive, labour is under constant pressure to become more "productive", which understandably leads far too often to haste, waste and unsatisfactory workmanship.

Labour input, however, is not the only item here that has been falling. Consider dimensional lumber. A so-called "two-by-four" board, for example, is now only 1½ inches by 3½ inches in size. Its cross-sectional area has thus fallen to 5¼ square inches (1½ in. x 3½ in.) from 8 square inches (2 in. x 4 in.). Similarly, the cross-sectional area of a "two-by-six" has fallen to 8¼ square inches (1½ in. x 5½ in.) from 12 square inches (2 in. x 6 in.). Under conventional economics, our lumber industry considers that most of the time the best thing to do with a good strong saw log is to pass a whirring saw blade through it so many times that each resulting board, after being planed, ends up being only 1½ inches thick. In other words, we use precious inanimate energy to deliberately and enormously weaken the very building materials that nature has freely given us! No wonder the disparaging term "stick-framing" has come to be used in certain circles to refer to the way we currently build the wooden framework of our houses. Modern stick-framing may be fast, but it certainly does not do justice to the trees we cut down. Nor is it energy-efficient when sawmill energy consumption is taken into account. Nor is it noted for its durability.

Under conventional economics, have we not deliberately sacrificed overall housing longevity in favour of higher "labour productivity" and greater so-called "economizing of materials" (i.e. economizing in the short term)?

Still on the subject of housing longevity, I turn now to the question of whether our intended high levels of "labour productivity" have perhaps been compromised by a decrease in the market value of the finished product. In other words, although today's new houses are definitely being built with fewer hours of human labour (i.e. with less human energy) than before, maybe their market value has fallen by an equivalent percentage. If so, then "labour productivity" would not have been climbing after all. Remember that output is measured in dollars, according to the definition of "labour productivity".

But is the market value of one of today's typical new houses really lower than that for some hypothetical new house that possesses a much longer life expectancy but is otherwise comparable? No one knows for sure, because few such houses are being built. Under conventional economics, however, there is reason to suppose that any extra market value in such houses is likely to be minimal.

Here is why. Because of our current fixation on money and on the interest that money seems to be able to earn indefinitely, we place a very low "present value" or "discounted value" on any asset whose benefit will be available to future generations but not to ourselves. We reason as follows. We imagine ourselves investing some money at, say, 5% compound annual interest. After one hundred years, a $1 investment would be worth approximately $130. (Yes, compound interest produces amazing long-term results!) We therefore conclude that any asset likely to be worth $130 in 2115 has a "present value" in 2015 of only $1. For a house getting built today, those numbers would translate into the notion that any of its features or aspects or qualities likely to have a value of say $130,000 in 2115 would have a "present value" in 2015 of only $1,000. Extra durability might well constitute such a feature. In other words, if a Charles Carter house and a conventional house were both built today, then in 2115 the former might conceivably have a market value that is $130,000 greater than that of the latter; but even so, the "present value" of the Charles Carter house in 2015 might be calculated to be only $1,000 greater than that of its conventional counterpart.

(Note that there is nothing wrong with the arithmetic here. But there is everything wrong, I and many others would argue, with the assumptions upon which the arithmetic is based.)

Accordingly, it is easy to imagine that under conventional economics the present market value of the extra durability of a house built to Charles Carter's standards might not amount to very much. That being the case, the market value of the output of the human labour allocated to the building of one of today's conventional houses is hardly compromised (if at all) by the unimpressive level of durability attained. Hence "labour productivity" levels, when calculated according to the accepted definition, are indisputably high at present. How reassuring to the ideology of perpetual economic growth and perpetual "labour productivity" growth! But how misleading!

Intelligent National Frugality (INF) economics

Under INF economics, our society will adopt a new set of attitudes in dealing with the problems mentioned in this chapter:

1. It will totally disregard all national "labour productivity" statistics (if anyone bothers to produce any) and all arguments based thereon.
2. It will focus on ensuring that each new generation of our population acquires high levels of personal skill and competence.
3. By granting to human energy the right of first refusal, as explained in Chapters 2 and 17, it will also focus on ensuring that the personal skill and competence acquired by each new generation get adequately applied to the tasks at hand. Assistance from non-human energy and from power tools and mechanized equipment will not necessarily be declined, but any acceptance will be judicious and discriminating.
4. By ignoring all the partial-truth arithmetic of "labour productivity", it will not allow such arithmetic to get in the way of intelligent approaches to health and safety.
5. It will place a very high value on durability, not just for housing but throughout the whole economy, because durability is so important both for energy efficiency and for frugality.

Conventional economics

In matters of "labour productivity" as in other matters, the trouble with conventional economics is that it focuses its attention almost exclusively on money and horsepower and short-term wealth. Too often it simply ignores the rest of what is going on. In particular, it ignores what is happening to the quality of our lives. It ignores most of the pollution, contamination, ugliness, stress, demoralization, chronic ill health, cynicism, community disintegration, economic insecurity, widespread wastefulness and general frustration that now characterize modern life. And because it mostly ignores such problems, it completely overlooks the possibility that their cause might lie primarily in our society's misguided efforts to constantly raise, at whatever cost, our "labour productivity".

As individuals, we can all feel the ongoing deterioration in the quality of our lives under conventional economics. So, as individuals, we all try to compensate in one way in particular. We try to compensate by striving frantically to increase our own personal hourly income. In other words, we try to compensate by continually raising, in one way or another, our own personal "labour productivity". But from the point of view of society as a whole, that is the equivalent of trying to outrun one's shadow, is it not? As this whole book argues, we need an alternative approach.

19

Free Enterprise versus Socialism

"In any case, both socialist and capitalist economic theory have apparently developed without taking into account the limited capacity of the biological capital represented by the ecosystem. As a result, neither system has as yet developed a means of accommodating its economic operation to environmental imperatives."

Barry Commoner (1971)

"The initial, wondrous promises of capitalism and corporate free enterprise have not always led to a careful and responsible management of the earth's natural resources and treasures. And the ills of communism led in many cases to disastrous pollution and wanton neglect of nature."

Helen Caldicott (1992)

In this chapter, I would like to touch on a number of different points all relating to free enterprise or to socialism or to both.

Neutrality

As I see it, Intelligent National Frugality (INF) economics does not take sides in the ongoing debate between free enterprise and socialism. It remains neutral because it is equally applicable to both systems, as well as to any intermediate position between the two.

Not only is it equally applicable, I would argue, but it, or something comparable, is also equally necessary. Neither free enterprise nor socialism can function satisfactorily with the skewed energy prices that characterize conventional economics. Ivan Illich expands on that point as follows: "A low energy policy allows for a wide choice of life styles and cultures. If, on the other hand, a society opts for high energy consumption, its social relations must be dictated by technocracy and will be equally distasteful whether labelled capitalist or socialist."

Most of us have been brought up to believe that the debate between free enterprise and socialism is a crucial one. At one time, moreover, the issue seemed intimately linked to the Cold War. Nowadays, fortunately, we are all able to relax a little and weigh the two systems on their respective intrinsic merits. We can each make up our own mind as to just exactly how much socialism, if any, we would like to see incorporated into an essentially capitalist system, or how much free enterprise, if any, we would like to see incorporated into an essentially socialist system. Unanimity on the subject is not required at all. Accordingly, in this book I do not try to persuade anyone to adopt a particular position on the matter.

Up until now, however, most of the debate has, in my view, not really been between free enterprise and socialism. Rather, it has been between a distorted version of free enterprise and an equally distorted version of socialism. I call the former "subsidized free enterprise" and the latter "subsidized socialism", with the subsidy—a massive one—being provided in both cases by future generations. As I see it, both sides in the debate have tended to accept without question the skewed price structure of conventional economics. In particular, both sides have tended to assume that, as regards free enterprise, only one version (namely, the conventional-economics version) was worth discussing. Both sides have marshaled their arguments accordingly. As a result, we really have not yet had much opportunity to focus our minds on the question of how satisfactorily or unsatisfactorily a system of free enterprise might perform under INF economics.

That last sentence would seem to reflect a basic bias on my part in favour of free enterprise. But it seems to me that nearly everyone shares that basic bias as a starting point. Who, for example, does not believe that a private citizen should be allowed to start up his or her own local hardware store? Virtually all of us, I think, believe in free enterprise at that level, just as virtually all of us believe in socialized police forces and socialized fire departments. Generally speaking, it is only when greater concentrations of economic power come into play that our opinions lose their unanimity.

Concentrations of economic power are likely to be much more dilute under INF economics than under conventional economics. After all, solar energy is inherently diffuse and does not lend itself to concentration with the same readiness that fossil fuels do. Hence as solar and solar-derived energy gradually displace fossil-fuel energy, economic power can be expected to become more and more decentralized, i.e. less and less concentrated. For that reason, I make the general assumption that under INF economics some sort of system of free enterprise—at least as regards small-scale business activity—will in most cases be given a chance to prove itself but will be replaced or modified or severely restricted in the event of unsatisfactory performance. Such an assumption is by no means the same thing as a strong conviction that the performance will in fact prove satisfactory. Moreover, free enterprise includes the freedom to set up and operate cooperatives and other non-profit organizations. And even under free enterprise, the rights and privileges of business corporations can and should be legislatively restricted so that they mesh with the values of society as a whole. In any case, the purpose of this book is to explain and advocate INF economics, not to promote free enterprise.

Some readers may perhaps feel that even under INF economics any and all regimes of free enterprise would be objectionable. I have no wish in this book to argue against their position. But I urge them, just as I urge all others, to pay close attention to the way in which energy prices, including the price of human energy, would get determined in whatever system they favour.

Rural landlordism

If a decision to combine INF economics with free enterprise does get made, there is one area in particular in which a stitch in time might well be worth nine. Rural landlordism strikes me as being a foreseeable danger that merits being nipped in the bud. Back in 1879, Henry George outlined the danger with one long sentence: "If one man owned all the land accessible to any community, he could, of course, demand any price or condition for its use that he saw fit; and, as long as his ownership was acknowledged, the other members of the community would have but death or emigration as the alternative to submission to his terms."

No one is suggesting that one single person might ever own all the farmland in a country as large as Canada. But in both large and small countries, it is easy to imagine the situation where substantial quantities of farmland are owned by absentee landlords and farmed by tenant families. Such landlordism generally has the effect of transferring large amounts of wealth from productive farming people to a parasitic class of landlords (in which class I would include parasitic corporate landlords and their shareholders).

Some theorists may feel than an investment in farmland by an absentee landlord is not essentially different from an investment in, say, 51% of the shares of a manufacturing company by an absentee shareholder. But in fact there is a world of difference. Farmland is an absolutely unique resource because food, like water, is an absolutely unique necessity. If our economic and land-ownership systems were blind to this uniqueness, then under INF economics there might well be a financial incentive for wealthy people to buy up large tracts of farmland, sit back and collect perpetual rents from tenant farmers. Preventing that from happening, however, is not difficult. The law could set limits, for example, to the amount of farmland that any one person or family is allowed to own. Moreover, the law could look right through the screen of corporate ownership or even follow the recommendation of Dave Henson and prohibit corporations from owning farmland in the first place.

There is no need to go into details here. If under INF economics we are looking forward to a renaissance of agrarianism and the small-scale family farm, as indeed I think we should be, then obviously we will want to ensure that our farming community does not escape the frying pan of agribusiness only to end up in the fire of rural landlordism.

Know thyself

I have already discussed my own general inclination to start out by giving some form of free enterprise the benefit of the doubt in most cases. But I would also like to mention a psychological trick that some of us play on ourselves in discussions relating to free enterprise. To explain the nature of this trick, let me use myself as an example.

Just because I say that I believe in free enterprise, and just because I believe that I believe in free enterprise, that does not necessarily mean that I really do believe in free enterprise. I may instead be deceiving myself. Self-deception is every bit as alive and well in the field of personal economic beliefs as it is in other areas.

If this whole idea sounds a bit strange, here are a couple of parallels.

Just because I say that I am always honest, and just because I believe that I am always honest, that does not necessarily mean that I am in fact always honest. I may in fact be both the world's greatest liar and the world's greatest self-deceiver.

Similarly, just because I say that I never behave in a racist manner, and just because I believe that I never behave in a racist manner, that does not necessarily mean that in fact I never do so behave. I may be as little aware of my own feelings and motivations as the man in the moon, and at the same time I may be highly practised in the art of racial discrimination.

In Wendell Berry's rural Kentucky novel *Jayber Crow*, the introspective title character is aware of his own psychology on this general point. He wonders, "Was I fooling myself? I know myself to be a man skilled in self-deception, and so maybe ... I ought to suppose that I was fooling myself."

If I do not in fact believe in free enterprise, then what do I really believe in?

ANSWER: I probably believe in maximizing my access to large quantities of electricity and mechanical horsepower and so on. I probably believe in maximizing society's access thereto as well, since that can only help maximize my own access.

Then why do I not openly say so, both to myself and to those around me?

ANSWER: Matters of the subconscious are not always easy to explain. Essentially, however, I think that people such as myself would rather see ourselves as holding a sincere philosophical belief in something as widely praised as free enterprise than see ourselves as hungering selfishly and shortsightedly after speed and power and instant gratification. In the words of Barbara Ward, we engage in "the self-deception with which interests are proclaimed as ideals."

How can one tell whether one really believes in free enterprise or whether the supposed belief is in fact an instance of self-deception?

ANSWER: By the direction that one takes when one reaches a fork in the road. Sometimes there is a clear divergence between (a) the principles of free enterprise and (b) ready access to speed and power. Nuclear power provides the most obvious example. As I have already mentioned, nuclear power does not get produced in the absence of large subsidies. The financial costs are simply too high, even under conventional economics. And yet nuclear power has a very strong attraction for a great many people. Accordingly, when we come to the fork in the road where one sign points towards free enterprise and the other towards nuclear power, we sometimes forget our original destination.

A second example, perhaps somewhat less black and white, can be seen in the modern aviation or air transportation industry. Writing in 2002 from a British point of view, John Whitelegg observes that modern aviation is a "massively subsidized industry" and is a "hugely significant source of ugliness in health, cultural, economic and environmental terms and yet it prospers through its success in creating a false consciousness of sophistication, excitement and pleasure." Whitelegg continues: "[Aviation] enjoys significant taxation advantages (no tax on fuel), airports are not regulated in terms of emissions in the same way that large industrial plants are, international aviation is not counted as part of any reduction target for greenhouse gas emissions, and large amounts of public cash are poured into supporting airports through new roads, motorway widening and new metro lines."

So at this second fork in the road, one sign points towards free enterprise while the other sign points towards inexpensive air travel and inexpensive air freight. At this second fork too, we sometimes forget our original destination. We turn instead towards speed and power. At the same time, we perform intellectual and verbal gymnastics in an effort to convince ourselves that no fork in the road ever existed in the first place. Such is the nature of self-deception.

Roger Terry goes a little bit farther: "Many people claim to believe in unhindered competition, but when push comes to shove, we discover that they'd actually prefer to have the government step in and ensure their success and prosperity, rather than having to 'earn' it (and possibly lose it) in the mercenary marketplace they extol ... So who really does believe in a totally free market? Perhaps no one."

One can lead a horse to water but one cannot make it drink. By the same token, you can point out to me an instance of self-deception on my part but you might have difficulty in persuading me to acknowledge that self-deception. We all have our pride, after all. That being the case, there is probably little to be gained by leveling accusations of self-deception at people such as myself, even in situations where its existence seems beyond question. A more fruitful approach might be to focus on the question of consistency, to which I now turn.

Consistency

In order to be intellectually persuasive, a set of beliefs must be internally consistent. One loses one's credibility if one tries to take two totally different stands on the same issue at more or less the same time. With that in mind, let us examine free enterprise. More specifically, let us compare the Intelligent National Frugality version of free enterprise (which I will call "INF free enterprise") with the version developed under conventional economics (which I will call "conventional free enterprise"). Let us do so under three different headings. And in each of the three cases, let us ask which version offers a more positive reflection of the basic values and ideals of theoretical free enterprise. The three headings are (a) non-wastefulness, (b) economic freedom and (c) innovation and wealth creation.

Let us begin with non-wastefulness. Free enterprise is prized for its reputed ability to allocate scarce resources efficiently. In other words, free enterprise is supposed to be less wasteful than any other economic system.

But when one looks around the modern industrialized world, most of which is being managed in accordance with conventional free enterprise, what does one see? One sees the two kinds of wastefulness (as described in the latter part of Chapter 17) everywhere, especially the structural kind of wastefulness: wasted human energy, wasted machinery, wasted electricity, wasted fossil fuels (notably through thousands of transportation absurdities), wasted heat, wasted air-conditioning and refrigeration, wasted wood, wasted metal, wasted newsprint, wasted packaging materials, wasted buildings (including dilapidated farm buildings), wasted interior space, wasted concrete, wasted railway lines, wasted sunshine, wasted water, wasted food.

The inconsistency between the theoretical non-wastefulness of free enterprise as a concept and the flagrant wastefulness of conventional free enterprise in the contemporary world would surely break poor old Adam Smith's heart.

I sometimes think that if one were to hire a group of first-rate economists with a good knowledge of human nature and if one were to give them instructions to design an economic system that maximized the overall amount of wastefulness in society as a whole, without actually causing the system to collapse overnight, then they would probably come up with something quite similar to conventional free enterprise.

In the matter of frugality and energy efficiency, which is to say in the matter of non-wastefulness, there would seem to be no contest between INF free enterprise and conventional free enterprise.

As a second cherished value associated with free enterprise, consider personal economic freedom. No one likes unnecessary government prying or unnecessary government regulation. But under conventional free enterprise and its taxation system, the government needs to be informed of every penny that you and I earn and of every penny that corporations earn. Detailed regulations specify which expenditures are deductible for taxation purposes and which are not. And an army of civil servants more or less spies on all of us in order to try to catch those who cheat.

Under INF free enterprise, by contrast, few (if any) of us will have to pay any income tax at all. We will therefore not have to worry about deductions for this or receipts for that. Under-the-table prices will be equal to over-the-table ones. Any dollar that I earn will be mine to keep and to spend as and when I wish. No government agency will need to know where it came from or where it went. That surely is a major part of what economic freedom is supposed to be all about. With regard to economic freedom too, then, INF free enterprise would seem to win the consistency competition by a wide margin.

My third heading relates to innovation and wealth creation. Both of these items supposedly receive far more stimulation under free enterprise than under any other system. Unquestionably, conventional free enterprise has encouraged innovation, so much so that our whole society now seems to regard innovation as being almost an end in itself. Equally unquestionably, conventional free enterprise has created large amounts of certain kinds of wealth. But as this whole book has been arguing, the cost of all this innovation and wealth creation has in many cases been far from negligible, especially when seen from the point of view of (a) the environment, (b) depreciation and depletion, (c) future generations and (d) those members of our own generation who have been unlucky enough to have fallen by the economic or environmental wayside.

There is in fact a basic difference between the kind of wealth produced under the two versions of free enterprise. Conventional free enterprise tends to focus on the short-term and on high levels of horsepower and energy consumption. INF free enterprise will focus on the long-term and on high levels of both energy efficiency and intelligent frugality. Conventional free enterprise focuses on installing an air-conditioning system in as many new cars as possible. INF free enterprise will focus on building a society that tries to minimize both the need for and the use of not just automotive air-conditioning but also private automobiles themselves.

In their book *What's the Economy For, Anyway?*, John de Graaf and David K. Batker provide an almost unbelievable example of the way in which conventional free enterprise can promote the production of short-term wealth at the expense of long-term wealth. In their example, a large American coal-mining company deliberately chose to mine only three of the ten seams of coal present in the location being mined. "Seven seams of valuable coal were completely wasted, treated as 'overburden,' dug up and plowed back into the pit, mixed with dirt, silt, sand, and rock." Why? Because the company was able to "make one percent more in profit by mining only the three best seams" and because the company wanted "to maximize returns this quarter."

A more common complaint about the nature of the wealth creation associated with conventional free enterprise can be seen in the following words of David Korten: "Even [the capitalist economy's] apparent capacity to create vast wealth is largely illusory, because though it is producing ever more sophisticated gadgets and diversions, it is destroying the life-support systems of the planet and the social fabric of society. It is therefore destroying our most important wealth."

So the consistency question here really depends upon whether theoretical free enterprise is supposed to create short-term wealth or long-term wealth. If, as one hopes, the answer is indeed long-term wealth, then here too INF free enterprise comes in a country mile ahead of conventional free enterprise.

I would also like to mention a fourth heading, but here I will confine myself to putting forward a simple question: As regards honesty and integrity and an ethical approach to the performance of whatever tasks are being undertaken, which of the two versions of free enterprise is likely to outperform the other? I leave it to the reader to arrive at his or her own answer to that question. But in Chapter 22 I will have more to say about ethics.

All in all, I suggest that conventional free enterprise fails the consistency test quite badly. It simply does not produce the kind of results that theoretical free enterprise is supposed to produce. By contrast, INF free enterprise would seem to pass the consistency test with flying colours. Admittedly, the success of

INF free enterprise is only theoretical at this juncture, since the system has not yet been tried anywhere. The failure of conventional free enterprise, however, can be seen concretely all around us.

This failure, when seen from the point of view of INF economics, presents no surprises at all. Conventional free enterprise involves the granting of huge government subsidies to nuclear and fossil-fuel energy. Those subsidies in turn cause huge price distortions. Given those huge subsidies and price distortions, there is no way in the world that conventional free enterprise could ever meet the performance standards associated with the theory of free enterprise.

Summary

Regardless of whether one favours free enterprise or socialism or some intermediate position between the two, there is in my view no justification at all for placing a large tax, or even a small one, on human energy and then using the resulting revenue to subsidize the price of nuclear and fossil-fuel energy. Such a policy can only be described as folly.

20

Advertising and Marketing

"A new consumer product must be introduced with a suitable advertising campaign to arouse an interest in it. The path for an expansion of output must be paved by a suitable expansion in the advertising budget. Outlays for the manufacturing of a product are not more important in the strategy of modern business enterprise than outlays for the manufacturing of demand for the product."

John Kenneth Galbraith (1958)

"Most of our long-held beliefs about money, wealth, productivity, and efficiency, and our notions of progress are rooted in immature, often infantile states of mind—easily manipulated by politicians and advertisers."

Hazel Henderson (1996)

Two criticisms and two subsidies

A great deal of the advertising that we see and hear all around us invites two strong criticisms. It wastes valuable resources (both directly and via the behaviour that it encourages) and it is deplorably manipulative. Not all advertising under every conceivable economic system deserves those two criticisms. But under conventional economics, much of our advertising has left legitimate merit far behind.

Let me begin with direct wastefulness. Consider, for example, the advertising technique whereby a whole page in a daily newspaper is purchased by an advertiser and then left totally blank or nearly so. Perhaps a corporate logo by itself will appear, or a brief phrase such as "COMING SOON" or "WE'VE BEEN LISTENING". As we all know, this mostly blank page constitutes only the initial portion, the attention-getting portion, of the advertisement. A page or two later comes the meat of the ad, occupying perhaps another full page or perhaps only half a page. Here we find whatever message it is that the advertiser wants us to stop and read.

Part, if not all, of the effectiveness of this particular advertising technique lies in its ability to give readers a tiny shock. "What?!" we say to ourselves, "Some company has spent thousands of dollars to buy that page of advertising and has left the whole page virtually blank! Incredible! I wonder what important announcement they want our attention for." We are indeed supposed to feel shocked at the amount of money involved. One doubts, however, that the advertiser wishes us to feel shocked at the waste of newsprint involved.

As seen from the point of view of Intelligent National Frugality (INF) economics, two major subsidies are at work here. One subsidy is general, while the other is more specific.

The general subsidy, as described throughout this book, applies to non-renewable energy, keeping the price of all such energy inordinately low. As a result, any advertiser purchasing one full page of a daily newspaper pays a subsidized low price for all the non-renewable energy allocated, directly or indirectly, to the production, transportation, printing and subsequent distribution of that one page of newsprint.

The second subsidy is created by the tax deductibility of advertising expenditures. Under conventional economics, business corporations are allowed to deduct advertising expenditures before calculating their corporate income tax payable. As a result, the after-tax cost of purchasing any advertising is usually only a fraction of the before-tax cost. The difference constitutes a significant subsidy.

When the two subsidies are added together, advertising becomes quite inexpensive, especially for large corporations earning large profits. And as the cost of advertising comes down, up goes the economic demand for it, i.e. up goes the economic demand by corporations to purchase advertising. That is why, under conventional economics, advertising bombards us from every corner. And in carrying out this bombardment, modern advertising consumes large quantities of real resources: electricity, paper and paper products, wood, metal, chemicals, glass, paint, ink and so on. Those large quantities all provide examples of the strong link between conventional economics and modern wastefulness.

Let us turn now to the manipulative aspects of modern advertising and modern marketing. Two aspects in particular are worth noting.

The first aspect concerns brand promotion. It is one thing for a business firm to promote its brand by means of advertisements that provide useful and truthful information. But it is quite another thing to advertise in such a way that psychological games are being played; the consumer is essentially being manipulated into preferring Brand X over other brands for all the wrong reasons. As Kenneth E. Boulding wrote as far back as the 1940s, "Most advertising, unfortunately, is devoted to an attempt to build up in the minds of the consumer irrational preferences for certain brands of goods."

The second aspect involves manipulating the public into feeling that they actually need or want or ought to have the product in question. John Kenneth Galbraith brought this aspect to everyone's attention in the words quoted at the beginning of this chapter.

Often, nowadays, those two manipulative aspects of modern advertising overlap. A firm's profits can be increased both by an expansion of its market share and by an expansion of the total size of the market. Manipulative advertising can help on both fronts.

A large soft drink company, for example, can generally earn sizable profits under conventional economics by plastering store-front after store-front, billboard after billboard, vending machine after vending machine and television screen after television screen with messages "reminding" us how natural and normal it supposedly is to build a good part of our lives around the consumption of soft drinks in general and around that company's brand of soft drink in particular.

Faced with such an onslaught, most of us end up by giving in, provided that we have enough money, or enough credit, to buy whatever it is that is being advertised. After all, no one enjoys feeling alienated. If, in my country, everyone else seems to drink this or smoke that or wash with these or cook with those or entertain this way or vacation that way, then I guess I too should do likewise. Large advertisers understand this psychology only too well, and they do their doubly subsidized best to exploit it.

In fact, whether it be on shiny new cars, on the latest clothing fashions or on expensive bank loans or credit-card loans, today's society tries very hard to get us all to spend our money as if there were no tomorrow. The manipulative impact of modern advertising and modern marketing appears to be cumulative. Here is Galbraith again: "More important still, the aggregate of all such persuasion affirms in the most powerful possible manner that happiness is the result of the possession and use of goods ..."

Hurry, hurry, hurry, my friends! Borrow, borrow, borrow! Spend, spend, spend! Buy, buy, buy! Smile, smile, smile! I will have more to say on manipulative behaviour in Chapter 22.

INF economics

Under INF economics, a bucket of ice-cold water will get poured over both the wasteful and the manipulative aspects of modern advertising and modern marketing. Each of the two subsidies mentioned above will get eliminated entirely. All business firms will then be left to pay by themselves the full unsubsidized cost of whatever advertising they wish to purchase. And whether it be for marketing purposes or for some other purpose, business firms will also have to pay the full unsubsidized cost of any entertainment, food, drink, travel and accommodation that they might wish to purchase, since no business expenses of any kind will be tax-deductible.

In addition to its direct effects, INF economics will also have major indirect effects on advertising. The whole national economy, as described in Chapter 3, will become more and more decentralized geographically and much smaller in scale. As a result, fewer and fewer brands of goods and services will be directed towards nationwide markets. Most will instead be directed towards modest-sized local markets. A corresponding decline can be expected in the amount of advertising purchased on behalf of national and international brands.

Fast-food franchises, for example, may disappear entirely, together with the advertising glue that keeps them in place. Some fast-food restaurants may well survive and prosper under INF economics. But their owners will have little incentive to spend much money on the purchase of a name-brand franchise. A virtuous circle will thus probably see national advertising and the whole franchising system each becoming weak and each further weakening the other.

As for television (including all its technological spinoffs), radio, newspapers and magazines, downsizing would seem to be the order of the INF day. By that, I mean less production and less consumption. At present, these industries generally receive most of their income in the form of advertising revenue. Such revenue seems likely to fall off considerably under INF economics. Or at the very least, such revenue will almost certainly decrease in relation to the cost of producing a newspaper page or producing and transmitting in one way or another an hour of television programming. Conceivably, direct revenue from readers and viewers might then rise sufficiently to take up the slack. But in most cases, here as elsewhere, demand will likely fall as price goes up. My guess is that INF economics will witness a substantial decrease in the annual tonnage of newsprint produced and consumed, in the annual total number of hours of television programming produced and transmitted and so on.

What about the Internet? Under INF economics, will the Internet be able to continue relying on advertising revenue in order to keep itself afloat in its current form? I make no prediction. On the one hand, the Internet has proven to be extremely popular under conventional economics. On the other hand, it consumes a great deal of highly subsidized electricity behind the scenes. At present, much of this cheap energy is paid for by advertisers. Once the INF taxation system is in place, however, the Internet's future will depend upon how willing we all are—advertisers as well as the Internet-using public—to pay the substantial energy costs involved.

Most people would agree, I think, that advertising requires a certain amount of regulation. On that point, neither conventional economics nor INF economics ties anyone's hands. Under either system, for example, society can severely restrict ugly outdoor advertising, ban all advertising from schools, restrict or prohibit tobacco advertising and impose strict rules regarding truth in advertising. As I have said before, however, society should first get the economics right, which will solve a great many problems all by itself. That done, society can use regulations, restrictions and prohibitions to tidy up any remaining loose ends. (I am speaking conceptually here. There is no reason to postpone any important regulation of advertising—or of anything else, for that matter—until the transition to INF economics is complete.)

Under any INF version of free enterprise, two things seem predictable. One is that a certain amount of advertising will continue to exist and to take various forms. The other is that advertising will consume far fewer real resources than at present. The same general frugality that will characterize the rest of the economy will characterize advertising as well.

As I see it, good advertising can play a useful role in society. Among other things, as Kenneth Boulding has himself acknowledged, it can help a business firm provide potential customers with the information they need in order to purchase the right product or the right service for the task at hand. INF economics will surely be judged, in part, on the basis of how well it culls out the anti-social aspects of modern advertising and modern marketing while leaving the more positive aspects untouched.

I am by no means the first person to aim at this goal. Charles Carter, in his short 1968 book, *Wealth: An Essay on the Purposes of Economics*, not only described and analyzed many of the more negative aspects of modern advertising and modern marketing, but also pondered the question of what kinds of government policies might improve the situation. "Any control," he wrote, "will clearly have to be financial," i.e. some sort of "a restriction on expenditure on the stimulation of consumer demand." He did not go so far as to suggest INF economics as a remedy. But he did suggest that, for taxation purposes, all deductions for in-country expenditures on advertising and marketing could be formally excluded from the calculation of business profits (unless such expenditures were covered by government-issued sliding-scale "Marketing Certificates", which he then went on to describe). In other words, he proposed that the government severely limit the second of the two subsidies (which I identified earlier in the present chapter) that promote the wastefulness, ugliness and manipulation associated with modern advertising and modern marketing.

I salute Charles Carter for being ahead of most of his peers on this matter.

21

Money Creation

"But if Galileo and Copernicus had lived to-day, and had upset the theories of the authorities regarding the nature of money rather than of the universe, they would have had far more difficulty in getting their new views impartially discussed than they had from the Medieval Schoolmen and the Courts of the Inquisition."

Frederick Soddy (1931)

"[Frederick] Soddy was a complex iconoclast often derided for dealing with what seemed to be disparate interests. What, it could be asked, could monetary reform possibly have to do with radiochemistry? But to make sense of Soddy, the question must rather be formulated in the other direction: What fundamental concern did Soddy have that enabled him to embrace holistically a variety of seemingly diverse activities? The answer can be given in a word: energy."

Thaddeus J. Trenn (1979)

An abundance of absurdities

Misery, it is said, loves company. So does absurdity. Under conventional economics, our absurd striving for perpetual economic growth is fuelled by the combination of (a) our foolish overuse of non-renewable energy and (b) our unthinking underuse and misuse of human energy. These energy absurdities in turn involve a whole host of more specific absurdities, many of which I have already mentioned: transportation absurdities, land-use absurdities, architectural absurdities, non-repairability absurdities, antibiotic absurdities and so on.

To that list, I would add the absurd way in which most of our money is currently created.

In this chapter, I would like to describe two different ways of creating money. One of the two is eminently reasonable but until recently has been little discussed and largely overlooked. The other, despite being in common use today, makes no sense at all. (Neither way involves the kind of money that is fully backed by something tangible such as gold or silver. But such backing has its own problems, which are beyond the scope of this book.)

From the point of view of Intelligent National Frugality (INF) economics, the first way of creating money that I am about to describe has several attractive features. Most importantly, this first way is able to function quite satisfactorily regardless of whether economic growth does or does not take place. As for the second way, it is highly unattractive for several reasons, including the key fact that, in the prolonged absence of economic growth, it would be socially disastrous.

Any country choosing to adopt INF economics will almost certainly prefer the first way, to which I now turn.

"Hundred Percent Money"

In its broad outlines, this first system is easy to understand because it is quite rational. Many people, including some economists, are intuitively attracted to it, even though conventional economics ignores it completely. Its name, if it has one (see below), is the "hundred percent money" system.

In the "hundred percent money" system, every single dollar (or euro or pound or whatever) of the currency in question comes

312

into existence in the same basic way. Conceptually speaking, what happens is that the central government of a newly created or newly independent country decides that, say, $10 billion worth of money is needed in order for the nation's economy to function properly. Accordingly, the government (via the central bank) prints up $10 billion worth of legal-tender bank notes. The government then gets all that brand new money into circulation by the simple expedient of spending it! The money might get spent on the salaries of civil servants, for example, or on a new parliament building or on pensions for the elderly. Thereafter, the money gets transferred in the normal way from buyer to seller, from employer to employee, from lender to borrower, from borrower back to lender, from donor to recipient, from taxpayer to government, from fined offender to fine-levying authority, from loser to finder and so on.

As for legal-tender base-metal coins, which for the sake of simplicity I will mostly skip over in this chapter, they come into existence in almost exactly the same way as legal-tender bank notes. The only difference is that nickels and dimes and loonies and so on are minted out of base metals such as copper and nickel, rather than starting out as sheets of paper (or a paper substitute) and then being transformed into $10 bills and $20 bills and so on by means of an elaborate printing system.

With $10 billion of newly created money now in circulation, economic activity proceeds in the normal way. But no more new money gets created unless and until some significant change in economic circumstances—such as an increase in population or an increase in the per capita size of the economy—suggests to the central bank that an increase in the country's money supply might be in order. The sole purpose of such an increase would be to facilitate normal commercial transactions in everyday life. To that end, very keen judgment on the part of the central bank is required. Too large an increase in the money supply could easily promote inflation. But too small an increase, or no increase at all, might choke off desirable economic activity. In any case, once the central bank has decided in favour of an increase, the printing presses would get started up again, another $1 billion worth, say, of legal-tender bank notes would get printed up, and then the government would spend all this brand new money into circulation, just as before.

Note the following six points:

1. As mentioned, government spending constitutes the specific act whereby all this newly created money gets placed into circulation.

2. Once created and spent into circulation, all this money becomes permanent money. It remains in existence until either destroyed (accidentally or deliberately, see below) or irretrievably lost.

3. When the government spends brand new money into circulation, the government obtains, in effect, a one-time free use of such money. To the extent that the government represents the whole of society, this one-time free use of brand new money can be considered as a sort of gift from society to itself. Ideally, there are no winners and no losers. The word "seigniorage", with variable spellings, is used to refer to this one-time financial benefit that accrues to money creators when they spend new money into circulation. During a certain period of European history, seigniorage accrued to the absolute monarch of the country in question, which explains the origin of the word.

4. Newly created paper money does not constitute newly created wealth. Society does not suddenly acquire $1 billion worth of additional real wealth just because that

314

much new money has been created. Money and real wealth should never be confused with each other, even though, whether in the hands of individuals, corporations or governments, money can easily be used to buy pre-existing real wealth. Provided, however, that the decision to create the $1 billion worth of new money was well-founded, the national economy will likely perform more satisfactorily for everyone than if the new money had not been created.

5. If the country's economy were to shrink in dollar-measured size, as might well occur under INF economics for all the right reasons, then the central bank would probably conclude that the country's money supply was now excessively large and likely to promote inflation. In that case, the government could carefully reduce the money supply by turning over some of its own cash to the central bank. The latter in turn would deliberately destroy the cash in question or at least lock it up in a secure vault until such time as new cash or replacement cash was needed. Once again, such destruction would neither benefit nor penalize any particular individual or entity.

6. For any country not starting out with the "hundred percent money" system, it is never too late to change horses in midstream, i.e. to switch over from our current system of money creation to the "hundred percent money" system. Upon completion of such a switch-over, every single dollar in the country's money supply will have become a permanent "hundred percent money" dollar.

Now, what about the name of this "hundred percent money" system? Back in 1935, the distinguished American economist Irving Fisher wrote a book titled *100% Money*. In that book, Fisher argued very strongly in favour of the "hundred percent money" system that I have just described. (Fisher was advocating this system within the context of conventional economics. As far as I know, his writings do not discuss the broad topic of energy taxation.) Nowadays, however, the term "hundred percent money" is used so seldom that few, if any, economics dictionaries even bother to include it as a term worth defining. They all mention Irving Fisher and they tell us about "Fisher's equation" and so on and they list some of Fisher's other book titles. But in general they omit any mention of the fact that Fisher not only directed his distinguished economic mind to "hundred percent money" but also wrote a whole book explaining it and advocating it.

This apparently puzzling omission is by no means confined to economics dictionaries. Consider, for example, John Kenneth Galbraith's 1975 book *Money: Whence It Came, Where It Went*. Galbraith there tells us quite a bit about Irving Fisher, including even a few personal details. We learn that Fisher had "sharp, good-humored eyes and an elegantly trimmed beard." And: "In the aftermath of the 1929 crash [Fisher] lost, according to his son, between eight and ten million dollars, a sizeable sum for even an economist." But we do not learn anything at all about Fisher's interest in, and writings on, "hundred percent money".

It appears that when mainstream economists, for whatever reason, do not wish to talk about something, they can be as uncommunicative as a cat. Ironically, Galbraith himself seems well aware of this dismaying tendency, for in the same book he also writes, "The young have always learned that Benjamin Franklin was the prophet of thrift and the exponent of scientific experiment. They have but rarely been told that he was the advocate of the use of the printing press for anything except the diffusion of knowledge." In my view, there is a close parallel between what, as alleged by Galbraith, rarely gets told to the young about Benjamin Franklin and what rarely gets told to anyone about Irving Fisher.

Fisher was neither the first nor the last person to severely criticize our current system of money creation and to advocate its replacement by the "hundred percent money" system. In particular, Frederick Soddy's name needs to be mentioned as well. Soddy was an outstanding British scientist who was awarded the Nobel Prize for Chemistry in 1921. He also thought about, and wrote about, economics. His most important economics book, written in 1926, is titled *Wealth, Virtual Wealth and Debt*. In that book, although he did not actually use the term "hundred percent money", Soddy advocated the same basic system of money creation that Irving Fisher was to write about almost a decade later. Incidentally, there are footnote and bibliography references to Soddy in Fisher's *100% Money*; so by 1935 Fisher was clearly familiar with Soddy's economic writings.

Mainstream economists ignored Soddy almost completely. They still do. Herman E. Daly sums up the situation as follows: "Perhaps the most intriguing thing about Soddy the economist is that he started his inquiry with a mind both highly intelligent and completely free from the preconceived paradigm of the orthodox economists, for whom he had an undisguised contempt. The contempt was mutual."

My reasons for according special attention to Frederick Soddy are essentially the same, I think, as the reasons of such writers as John B. Cobb, Jr., Herman E. Daly, Joshua Farley, Gerald Foley, Linda Merricks, Andrew Nikiforuk and Thaddeus J. Trenn for doing so. Not only did Soddy advocate the "hundred percent money" system (without ever using that term), but he did so because he felt strongly that our existing system of money creation is intimately linked with our strange fixation on perpetual economic growth and with our strange reluctance to look energy squarely in the face. I personally thus feel drawn more closely to Soddy's economic work than to Irving Fisher's work. Fisher's *100% Money*, however, has the virtue of spelling out in clear language and considerable detail how the "hundred percent money" system would actually work.

The fractional reserve system

I turn now to the second of the two systems of money creation that I wish to discuss. For reasons that I will mention in a moment, it is called the fractional reserve system.

Note first of all the following three preliminary points:

1. Most, if not all, modern countries are using the fractional reserve system at the present time. Hence it is described in dozens of different economics textbooks. According to an important article by Michael McLeay, Amar Radia and Ryland Thomas, however, some economics textbooks do not describe the working of the fractional reserve system correctly.
2. The fractional reserve system is somewhat difficult to understand, partly because it is rather complex, partly because it is not always described correctly (as mentioned in the previous point) and partly because, in the view of a good many people including myself, it is preposterous.
3. The fractional reserve system is extremely unlikely to be retained under INF economics.

Now for the nitty-gritty. The key fact concerning the fractional reserve system is that it combines two economic activities into a single process, even though, conceptually speaking, the two activities are quite distinct from each other. One of the two activities is money creation. The other activity is the lending of money by commercial banks. Why combine money creation with money lending? Don't ask me, I just work here! I would be the last person to describe this system as being logical or sensible. Its sole importance arises from the fact that it is in actual use in a great many countries around the world.

Under the fractional reserve system, only a small portion of the country's money supply is created in accordance with the principles of "hundred percent money", as described above. This small portion consists of all the legal-tender coins and legal-tender bank notes that are in circulation: in wallets, purses, pockets, cash registers, automated teller machines, bank vaults (but see below), store safes, armoured cash-delivery trucks, cookie jars, mattresses, gangster suitcases and so on. But most of the money supply is actually created by the commercial banks. Out of thin air!

Admittedly, any and all money created under the "hundred percent money" system is also created out of thin air. But no one pretends otherwise. That is what seigniorage is all about. Under our fractional reserve system, however, the commercial banks do not exactly fall over backwards trying to make sure that we all understand where our money supply comes from. If we, the public, want to believe that the commercial banks do not create money but rather merely lend out pre-existing money that their customers have previously deposited, the banks' attitude tends to be: "Let sleeping dogs lie!" In recent years, happily, more and more slumbering dogs are beginning to wake up.

(For those interested, I should perhaps note here that, in describing the fractional reserve system, economics textbooks generally exclude the cash stored in bank vaults from calculations of a country's total money supply. Such exclusion merely avoids double counting. If I have $100 in my bank account, for example, and if the bank has only $10 in cash in its vaults to "stand behind" my account, then the total amount of money—legal-tender money plus non-legal-tender money—in existence here is $100,

319

not $110. It would be an arithmetical error to count the $10 cash twice: once as cash in the vault and once as part of my $100 account.)

Let us now look more closely at a typical bank loan granted under the fractional reserve system. When a commercial bank lends, say, $5,000 to a customer, the bank does not normally hand over $5,000 in cash. If it did, then no money creation would be involved. But normally, the bank merely enters a credit of $5,000 in the customer's chequing account in that particular bank. No cash changes hands. Instead, strange as it may seem, $5,000 worth of brand new money gets created out of thin air by the lending bank. If I may repeat myself, the bank does not lend out pre-existing money that has previously been deposited in that bank by someone else.

As stated, this brand new $5,000 does not take the form of legal-tender cash. Nevertheless, it qualifies as perfectly real money. It takes the form of "money in the bank". Using such money together with a cheque book or a debit card or an electronic device, holders of bank accounts can pay bills, make purchases, pay the rent, pay off loans and so on. A stroke of the pen, so to speak, or a key-stroke of the computer is all that is required in order for this brand new $5,000 to be created. Instantly, the nation's total money supply gets increased by $5,000.

By the same token, whenever a bank customer uses a cheque or a computer to repay, say, $100 of principal on a bank loan, the act of repayment causes the country's overall money supply to decrease by $100. The $100 of bank-created money comes back out of the customer's bank account and disappears into thin air once again. Now you see it, now you don't! Money that is created by bank lending has only a temporary existence. It is thus very different from "hundred percent" money. And in order for our bank-created money supply to remain adequately large, the various kinds of borrowers (institutional, corporate and individual) in our society have to keep on borrowing money from our commercial banks indefinitely. Only thus does each dollar of disappearing money get replaced by a brand new dollar having, like its predecessor, a distinctly temporary existence.

Impermanence does not constitute the only major difference between the two systems of money creation. Temporary money created by a commercial bank earns interest for its creator (i.e. for the bank doing the lending) throughout the whole time that the money is in existence, provided of course that the borrower eventually pays off not just the principal but also the accrued interest. By contrast, legal-tender bank notes do not earn interest for anyone, at least not simply by virtue of their existence. They can only earn interest for a lender when the lender lends some pre-existing bank notes to a borrower. And once such a loan has been repaid with interest, the bank notes in question become non-interest-earning once again. Put another way, a ten-dollar bill stuffed into a mattress will not earn any interest for anyone no matter how long it is left there. But a dollar created by a commercial bank will normally earn interest for its creator from the day that it is created until the day that it is extinguished.

The notion that bank-created money disappears with each repayment of principal may seem at odds with the facts. If my employer, for example, uses money borrowed from a bank to pay my wages, and if I then deposit that money in my own bank account, will the money disappear when my employer pays back his or her loan? Obviously, none of the money in my account will disappear. But the money (transmitted by cheque or electronically) with which my employer pays back the loan will disappear. Hence the country's total money supply will shrink by the amount of the loan repayment.

One might think that under the fractional reserve system the commercial banks could easily get careless with their lending and thereby flood the whole national economy with excessive amounts of newly created money. And that does happen on occasion, especially in times of economic boom. There are two kinds of constraints, however, that lessen—but do not eliminate—such likelihood. One of the two, while formal, is in many cases ineffectual. The other, while more indirect, works most of the time.

The formal constraint lies in the fact that in many jurisdictions each commercial bank is required by law to hold cash reserves that are at least equal in value to a certain specified percentage of the total amount of money currently on deposit in the bank in question. But the specified percentage is often extremely low. Hence this legal constraint seldom accomplishes much.

Suppose, for example, that I have a chequing account containing a balance of $1,000 in Bank X. Suppose further that Bank X is required by law to hold in its cash reserves an amount of money equal to at least 10% of the value of the current balance in all its chequing accounts. This means that, for my account alone, Bank X must keep a reserve of at least $100 cash in its bank vaults (or, what amounts to the same thing, it must keep at least $100 on deposit in its own account at the central bank of the country in question): not the full $1,000, just $100 in legal-tender cash. That is why the system is called the fractional reserve system. (Whether it is called the fractional reserve system of banking or the fractional reserve system of money creation, it is one and the same thing.)

The second and more effective kind of constraint against excessive lending by the commercial banks stems from commercial realities, as McLeay, Radia and Thomas have pointed out. I will single out here one such reality in particular. Not every would-be borrower is creditworthy. If Bank X lends out an excessive amount of money and overlooks the inadequate creditworthiness of too many of its borrowers, many of the loans will never get repaid in full with interest. The profitability of Bank X will suffer accordingly. In severe cases, the bank may even become insolvent, to the chagrin of its shareholders.

In the modern world, what unfortunately weakens this indirect kind of constraint is that sometimes an insolvent bank will be judged by the national government to be too big to be allowed to fail. The repercussions of the bank's failure might prove disastrous for the whole national economy and possibly for the global economy. In such cases, the government is likely to bail out the insolvent bank at taxpayer expense. The bank may thus survive to sail again on calmer seas, regardless of its past imprudence.

Given the weaknesses of the above two kinds of constraints, an obvious solution would be to raise the reserve requirements all the way up to 100%. But that would take us right back to the "hundred percent money" system advocated by Frederick Soddy and Irving Fisher and many others. Bank lending would then consist exclusively of lending out pre-existing permanent money. Bank lending would be completely divorced from money creation. There would no longer be any fractional reserve system at all!

There are in fact several drawbacks to the fractional reserve system. Over the years, many different people have articulated these drawbacks, including many people not necessarily sympathetic towards the energy taxation ideas of INF economics.

One of the most serious drawbacks is that the fractional reserve system sometimes allows the overall size of the country's money supply to fluctuate wildly. Because it consists mostly of temporary money, the total money supply can expand excessively during economic booms, as already mentioned, and it can shrink excessively during major recessions. Such wild fluctuations would be unlikely under the "hundred percent money" system where all money is of permanent duration. As I see it, money created by the fractional reserve system comes and goes like the wind, whereas "hundred percent money" more closely resembles the sky itself.

A second drawback is that, for no good reason, the fractional reserve system deprives the central government, and thus society as a whole, of the opportunity to benefit from a one-time free use of a substantial amount of money, as explained earlier.

A third drawback, related to the second, concerns fairness. Why should the commercial banks be allowed to charge substantial interest on the loans they grant, given that the money they lend is created out of thin air? Are we not simply presenting the banks, their senior officers and their shareholders with a gravy train? After all, not much effort is required in order to wave a magic wand and say, "Presto!" The only real work done by the banks in this matter consists (a) of doing the bookkeeping and (b) of being careful (not always, as we now know) to select borrowers who seem likely to repay their loans in full with interest. A little bit of preliminary prudence predictably produces prodigious pecuniary profits from profoundly peculiar practices!

A fourth drawback is that, to put it bluntly, there seem to be no theoretical arguments at all in favour of the fractional reserve system. Economics textbooks try to tell us how the system works, but not what its merits are. They cannot discuss its merits, of course, unless they are prepared to compare it with the obvious alternative, namely the "hundred percent money" system. But the latter has become, as discussed above, essentially unmentionable in mainstream economic circles. Hence textbook writers are left with no choice but to skip over the whole question of what merits, if any, the fractional reserve system might possess. In response to such silence, it is difficult to prevent one's mind from inquiring as to which group of people might be likely to benefit financially from the existence of a taboo in this matter.

Some writers choose stronger language than mine in dealing with the general ideas of the previous paragraph. William F. Hixson, for example, writing from an American perspective in 1997, puts forward the following question: "What is the proper percentage of all created money that should be Government-Created Money [as distinct from Bank-Created Money]?" Hixson then makes the following comment: "I have no hesitancy in asserting that there is a conspiracy of silence regarding this question. There is no alternative to this conclusion except to believe that everyone in government is abysmally ignorant about money and, although this may be true of a majority of officials, there is a sizeable minority that cannot profess ignorance and thus must be part of the conspiracy of silence." For the record, Hixson's answer to his own question is "that the government should create 100 percent of our money."

Irving Fisher sums up the arguments against the fractional reserve system in a single low-key sentence: "If our bankers wish to retain the strictly banking function—loaning—which they can perform better than the Government, they should be ready to give back the strictly monetary function which they cannot perform as well as the Government."

INF economics

As I stated at the beginning of this chapter, I feel fairly confident that INF economics will end up allying itself with the "hundred percent money" system of money creation. After all, the four abovementioned drawbacks of the fractional reserve system are not easily forgiven. Moreover, a fifth drawback exists as well. And in the context of the kind of society that I am proposing in this book, the fifth drawback is the most serious of all.

The problem here is that, in the absence of ongoing economic growth, the fractional reserve system of money creation would eventually become unworkable. Suppose, for example, that 5% of Canada's total money supply were to consist of "hundred percent money", while the remaining 95% consisted of fractional reserve money. Under such an arrangement, 95% of Canada's total money supply would constantly be earning interest for the commercial banks, month in, month out, year in, year out, decade in, decade out, century in, century out. In the absence (a) of perpetual economic growth and (b) of a corresponding need for perpetual growth in the money supply, where would the money to pay all this interest come from? Would not the banks and their shareholders become richer and richer while sucking the financial lifeblood out of all those of us unlucky enough to have to borrow money from the banks for whatever reason? I cannot imagine that under INF economics any fair-minded person would ever contemplate such an arrangement.

The only reason why the fractional reserve system works even half-satisfactorily under conventional economics has to do with economic growth. As the economy grows, more money is needed for normal financial transactions. The banks oblige by lending out larger and larger total amounts of money, some of which then gets used to pay both the capital and the interest on previously made bank loans. Herman E. Daley and Joshua Farley sum the situation up as follows: "Thus, a requirement for growth (or else inflation) is built into the very existence of our money supply."

A ridiculous taboo

Like the proverbial cat, some ideological taboos seem to have nine lives. They just refuse to die. As an example, consider the ridiculous but long-lasting taboo—in mainstream economic circles—against any meaningful discussion of the "hundred percent money" system. That particular taboo, however, may finally be nearing its end. If so, it will have been given an extra push towards oblivion by a 2014 book titled *Money: The Unauthorized Biography* by Felix Martin.

In that book, incidentally, Martin says nothing about energy, so I do not know what views he might hold about INF economics. But he has much to say (a) about the nature of money and (b) about the history of money and banking from ancient times right up to and including the global financial crisis of the early 2000s. "What is needed," he observes in a later chapter, "is reform targeted at the fundamental structure of the banking system, rather than at the behaviour of the bankers within it."

Martin then suggests three principles as starting points:

1. "What is required is a closer match—not a perfect one—between the costs and benefits that taxpayers, bankers, and their investors are at risk of bearing." In other words, let us have no more of those totally unfair situations where heads I win and tails you lose, i.e. where the banking community gets most of the benefits in good times and where taxpayers suffer most of the losses in bad times.
2. "[A]ny redesign must maximise the room for monetary policy" and preserve "the escape valve of a flexible monetary standard." In other words, there needs to be a direct mechanism by which the government (or some government-controlled institution such as a central bank) can—in the public interest—increase or decrease, as it

sees fit, the total amount of money in circulation, rather than having to sit back and let market forces determine what happens.

3. "The trick is to set as few rules as possible and police them rigorously, while setting private initiative and innovation free for the rest." In other words, free enterprise is desirable in most areas of the economy, but not as regards normal money creation. Moreover, "attempting to supervise the financial sector is a fool's errand." Rather than vaguely defined and questionably effective supervision, the situation requires a small number of crystal-clear rules all of which get enforced with ferocious strictness.

Next, Martin asks and answers, with reasons, the following question: "Is there any realistic proposal for banking reform that answers to this daunting job description?" His answer: "Irving Fisher's "famous [1935] proposal with the inspiring title *100% Money*." (I chuckled when I read the word "famous", knowing how studiously Fisher's proposal has been ignored until now.)

Martin also comments, "Today, under the banner of 'Narrow Banking,' [Fisher's proposal] is being advocated once again by some of the world's leading regulatory economists."

As I say, the ridiculous taboo on discussing this matter may finally be nearing its end. Good riddance!

Three oddballs and a good-natured twosome

At the present time, we have set up shop for a bizarre threesome consisting of (a) highly subsidized non-renewable energy, (b) the ideology of perpetual economic growth, and (c) money creation by means of the fractional reserve system of banking. I suggest that we chase all three oddballs away and that we invite in, as their replacement, the good-natured twosome of INF economics and "hundred percent money".

Part 5: Conclusion

22

Ethics and Economics

"Then and now, we are warned against the arrogant misuse of power. Then and now, we are reminded that to demand too much of our existence, whether in consumption or in power, risks the destruction of its physical and moral base. The old philosophers and the new scientists ... are beginning to speak the same language, plead for the same modesty and concern, warn of the moral dangers of overweening confidence, and ask for its opposite—respect for living things, especially the smallest, and cooperation, not exploitation, as the pattern of existence."

Barbara Ward (1976)

"The most important, difficult, and neglected questions of energy strategy are not mainly technical or economic but rather social and ethical."

Amory Lovins (1977)

Ideologies

With regard to ethics and morality, it is not surprising that economic and political ideologies tend to stumble badly. The problem is that ideologies (as defined in Chapter 13 above) are by nature closed-minded and dogmatic, whereas an ethical approach depends first and foremost on the continual asking of awkward and upsetting questions.

We can see this quite readily when we separate conventional economics into its scientific component and its ideological component. Economic science, if left to its own devices, has no difficulty in making room for ethical considerations. After all, the essence of the scientific approach to economics boils down to a constant willingness to expand the number of economic scenarios that need to be studied and understood. There would be no conceivable excuse for excluding from serious study those economic scenarios that place a high value on ethical considerations. The scientific approach is inclusive, not exclusive.

By contrast, the ideological component of conventional economics has already made up its mind as to what is good for us. Perpetual economic growth is good for us. Heavy taxation of human energy is good for us. Light taxation of fossil-fuel energy is good for us. Nuclear power needs to be affordable and therefore needs to be subsidized. Frugality is not worth discussing. And so on. All the key questions having already been answered, there is no reason to go back and ask them again. Conventional economics thus becomes exclusively a matter of expertise, with ethics being treated as irrelevant.

Power and corruption

In this section, I would like to expand on a point that I put forward in the Introduction.

With most ideologies, including the ideological component of conventional economics, we do not have to look very far in order to see what is going on. One word sums up both (a) the reason why the ideology in question seems, for a while, to be so appealing and (b) the reason why that same ideology inevitably drags ethical standards downwards. That one word is "power". As the words of Lord Acton (written in 1887) tell us, "Power tends to corrupt, and absolute power corrupts absolutely." We humans relish power, especially we males. We are instinctively attracted to it. But if we get our hands on too much of it, then our judgment tends to become corrupted and the ethical level of our behaviour tends to plunge.

We are all familiar with the link between power and corruption, at least in those situations where the power is of the kind associated primarily with political power and dictatorship. But the link is often overlooked in those situations where the power is of the kind associated with gasoline, diesel fuel, jet fuel and electricity. Nevertheless, I and others would argue that possession of excessive power over our physical surroundings can be almost as corrupting as possession of excessive power over other human beings. Indeed, under conventional economics have we not been slowly but surely corrupting ourselves by availing ourselves of too much mechanical horsepower, too much gasoline, too much electricity and so on? Back in 1974, Ivan Illich made that very point: "[B]eyond a certain threshold, mechanical power corrupts."

Alan Foljambe, drawing on his own cabinetmaking experience, observed that Lord Acton "was referring to politics but could as easily have been referring to tools," i.e. to power tools. Injudicious use of the latter can quickly corrupt one's determination to pursue high craftsmanship standards. In Foljambe's words, "This is what is lost when you have 240 volts to hurl at a recalcitrant slab of oak."

In the physical sciences, power is defined as the amount of energy that gets used or released in a given time period. And in ordinary life, the power that I am referring to in this chapter is much the same. Under conventional economics, we are corrupting ourselves by consuming too much energy each hour, each day, each month and each year. We are consuming too much energy too fast. We are availing ourselves of too much physical power.

Here is another (albeit fictional) example. Last Tuesday, I was scheduled for a very important job interview at 10:30 am at a location thirty minutes' drive from my home. Giving myself plenty of extra time, I planned to set off in my car at 9:30 am. But just after breakfast that morning, my wife and I got into a heated discussion over a personal matter. Time flew. At 10 o'clock, I grabbed the car keys, hurried out to my car and tore downtown, tires screeching, engine roaring, horn blasting, brakes furiously converting the vehicle's kinetic energy into gobs of waste heat at every stop. I bellowed and cursed at half a dozen motorists and pedestrians along the way. By good luck, and only by good luck, I arrived safely and without having harmed anyone.

Power does that. It corrupts our values and our judgment. As far as I can see, our only effective recourse is to refuse to accept excessive amounts of power in the first place. If I say, "I am an incorruptible person, so you may go ahead and entrust me with as much power as you wish," then I am almost certainly lacking in self-knowledge. What I need to say instead is: "Look, I am just as corruptible as anyone else. Do not give me any more power than is reasonable for the job at hand."

Note that if, in the above example, the only reasonable way for me to get to my appointment had been by city bus, then my outrageous behaviour could not have occurred. Or if my car had been very low-powered, then even outrageous driving on my part would not have created quite so much danger for others. In general, the degree of resulting corruption tends to be proportional to the amount of excessive power involved.

Here is still another example of the way in which physical power, when combined with conventional economics, can corrupt the judgment of an individual. I have taken this example from a first-person account by John Cameron who tells the story of how one day, in the mountains of Montana, "I winced as I drove my pickup truck behind a bulldozer gouging its way up a narrow old track, felling the fir and aspen trees that crowded the sides." He winced because he was the very geologist who had ordered this bulldozer work to be done. "I drove back to my cabin that night feeling like a criminal. The damage done by the bulldozer in widening and extending the track was insignificant compared with what would happen if the drilling were successful and an open pit mine were developed on site ... It was ironic that the person [Cameron himself] who had perhaps walked this section of the range more thoroughly than anyone else in recent times, who so appreciated the way it was, would be the one to inflict this wound and maybe many more ... How was I able to maintain selective blindness about what I was doing for so long?"

I salute John Cameron for his honesty and his courage. In terms of my argument about power and corruption, Cameron accepted, for a time, enough decision-making power to inflict ugliness and destruction on a particular landscape that he cherished: the beginnings of a classic case of spiritual corruption. But I will let Cameron have the last word: "The drill rigs didn't strike any silver or gold that season, fortunately. Painful though that time was, it brought me to face some hard questions and set me on a path back towards more congruent expressions of my love of the land, something that I now consider vastly more precious than gold."

Incidentally, Intelligent National Frugality (INF) economics by itself will not necessarily halt all mining for gold and silver. But a person or corporation or bank will have to want those metals pretty badly in order to be prepared to pay the hefty tax on all the fossil-fuel energy consumed in finding, mining, transporting and refining the mineral ore. Moreover, as I have said before, strict government regulations can always be imposed so as to provide as much additional environmental protection as is felt desirable.

When one thinks of all the corruption of judgment that takes place under power-crazed (is that too strong a term?) conventional economics, one cannot help but conclude that modern industrial society is treating future generations in a highly unethical manner. We take a great many selfish environmental risks at their expense, notably with regard to (a) radioactive nuclear wastes, (b) atmospheric and climatic stability, (c) chemical pollution and (d) destruction of the habitat of a great many living organisms. We are also rapidly depleting—far faster than our own legitimate needs could ever justify—the Earth's stock of various non-renewable resources, including fossil-fuel resources.

Such "I win, you lose" selfishness is unethical in the extreme. No wonder some of us try to cover up our embarrassment by asking half-seriously, "What has posterity ever done for me?" (One doubts that that question gets pondered by very many women. If it did, and if the pondering were to continue for long, then the human race might be in serious trouble indeed.)

The chain of cause and effect seems quite clear here. Making use of large amounts of fossil-fuel energy in particular, we have granted ourselves an enormous amount of power over our physical environment. Such power has corrupted our judgment. As a result, we have used much of that power selfishly and unwisely. Nature's time scales being what they are, future generations will probably have to pay for, or suffer from, most of whatever environmental damage ensues.

The economic wayside

Conventional economics tends to condemn a certain percentage of people in our society to years of degrading poverty, alienation and unproductiveness. Admittedly, not every case of such ills can be attributed solely to conventional economics. But

338

when one looks at the number of decent jobs and self-employment opportunities that have been unnecessarily confiscated from human energy and allocated instead to nuclear and fossil-fuel energy, the culpability of conventional economics would seem to be beyond question. Moreover, the extent of the culpability only increases when one remembers that innocent children often share in any unjust deprivations inflicted on their parents.

Here is Noam Chomsky on this matter: "The fact is that people's lives are being destroyed on an enormous scale through unemployment alone. Meanwhile, everywhere you turn you find work that these people would be delighted to do if they had a chance. Work that would be highly beneficial both for them and their communities. But here you have to be a little careful. It would be beneficial to people, but it would be harmful to the economy, in the technical sense. And that's a very important distinction to learn. All of this is a brief way of saying that the economic system is a catastrophic failure."

It is impossible to overstress the ethical difference between (a) an economic system that offers to human beings the right of first refusal as regards opportunities to engage in productive work and (b) a system that, in effect, offers that same right of first refusal to nuclear and fossil-fuel energy instead. (Nuclear and fossil-fuel energy can be thought of as "refusing" a job offer in those situations where "acceptance" is impossible because the requisite technology has not yet been adequately developed.)

Once again, the chain of cause and effect is not difficult to follow. We have granted ourselves (as a society) an enormous amount of physical power in order to provide ourselves with an unprecedented selection of goods and services ranging all the way from basic necessities to ridiculous absurdities. Such power has substantially corrupted our judgment. As a result of this corruption, we have difficulty in perceiving the structural wastefulness (see Chapter 17) that is built into conventional economics. Nor do we readily acknowledge, even to ourselves, the absence of any meaningful link between (a) very high levels of resource consumption (direct and indirect) and (b) high levels of human satisfaction and happiness. And the fact that significant numbers of human beings get left by the economic wayside leaves us cold. Power does indeed corrupt.

The ethical wayside

Condemning a certain percentage of our own people to the economic wayside is not the whole of the ethical story here. Under conventional economics, many of the rest of us are only able to avoid that economic wayside by allowing ourselves to fall by the ethical wayside. Indeed, many of us find ourselves earning our living by engaging, directly or indirectly, in highly manipulative forms of commercial behaviour. All too often, the thrust of our commercial effort is towards selling some product or service that we know full well to be inappropriate for the intended buyer or buyers. Sometimes the item is inherently unsafe or harmful to human health, as in the case of cigarettes. Sometimes the item is vastly overpriced. Sometimes it is absurdly wasteful or environmentally suspect. Sometimes its quality is unconscionably poor. Sometimes several of the above apply. Nevertheless, we push on with our salesmanship. We cannot bring ourselves to treat our customers selflessly rather than manipulatively. We cannot bear to forgo the income that success in selling will bring us.

The actual salesperson on the floor does not always bear the greatest ethical responsibility in these situations. In many cases, the sales staff are merely doing what is expected of them by their employers. The latter, together with such professional advisers as lawyers, accountants and advertising executives, must usually accept the lion's share of the responsibility. But once again the real villain, in my view, is conventional economics.

Conventional economics generates manipulative salesmanship in the same way that it generates joblessness: by confiscating from human energy far too many legitimate productive opportunities and then allocating those opportunities to nuclear and fossil-fuel energy. Other things being equal, most people would prefer to be doing work that is ethical, I believe. But when decently remunerated ethical work is in short supply, in which direction is a person to turn? Towards crime? Towards unemployment? Towards work that is lawful but ethically tainted?

I realize that a certain amount of manipulative salesmanship doubtless existed in ancient times and will doubtless continue even under INF economics. But in my view conventional economics strongly aids and abets such salesmanship. The whole of society suffers as a result.

Health and safety

Conventional economics is also open to severe ethical criticism for the many ways in which it unnecessarily endangers our health and our safety. Here too, in my view, our judgment and behaviour have been corrupted by the excessive quantity of physical power that we have allocated to ourselves. I will let my examples speak for themselves.

First, consider transportation accidents. Such accidents have no doubt been a major cause of death and injury ever since our ancestors first learned to ride on horseback. But in no way does the intrinsic danger of transportation constitute an excuse for the

341

present frequency of transportation-related deaths and injuries in wealthy places like North America.

Three factors in particular account for the ongoing tragedy. All three are directly related to conventional economics. One is the unprecedented total distance that most of us find ourselves travelling each year. Another is the tremendous tonnage of freight that gets moved around each year and the total distance involved. And the third and most important is the vast number of independently guided high-speed and high-powered motor vehicles that we make use of for the purpose of moving ourselves and much of our freight around, winter and summer. All these numbers are so excessively high that, statistically speaking, the number of transportation-related deaths and injuries cannot help but be high as well. And once again the fact that many of the victims are children only compounds the tragedy.

As a second example, consider industrial and construction accidents. Very often, the best safety device in the world consists of an extra pair of hands, an extra pair of eyes, an extra pair of ears. But under conventional economics, this is the one safety device that all too often ends up being prohibitively expensive. So all too often we dispense with the "extra" safety person and, predictably, the accident rate turns out to be shockingly high. Does the evening newscast not tell us with mind-numbing frequency that the truck-driver or the equipment-operator "didn't see" the person who ended up being crushed? And in the terrible 2013 runaway train accident in Lac-Mégantic, Quebec, not a single person was in the locomotive or anywhere else on the train.

Thirdly, consider high-stress occupations. They are notorious for sapping people's health. But the high levels of stress involved can often be traced mainly to under-staffing: too much work having to be done, and too many decisions having to be made, by too few people in too short a space of time. Under-staffing makes no logical sense at all when many well-qualified persons remain involuntarily unemployed or underemployed. Only a poorly designed system such as conventional economics would ever promote unnecessary under-staffing and its related stress and health problems.

Fourthly, consider synthetic chemicals. Many such chemicals are known to be, or are suspected of being, capable of causing cancer and/or other serious health problems. But the implications of that fact seem to overwhelm us. Like the proverbial drowsy frog in the slowly warming pot of water on the stove, we cannot seem to summon up the necessary determination to take decisive action, not even when it is our own lives that may be at stake.

Back in 1962, when Rachel Carson's classic *Silent Spring* was first published, many readers of that exposé of chemical harm could no doubt truthfully claim, "I didn't know." But more than fifty years have elapsed since then. Admittedly, we have now banned (completely or partially) some of the worst offenders. Under conventional economics, however, we have in general applied the same assiduity in promoting our chemical industry as in promoting our automobile industry, our aviation industry, our communications industry and so on.

To illustrate such assiduity, here are some numbers presented by Theo Colburn, Dianne Dumanoski and John Peterson Myers in their 1996 book *Our Stolen Future*: "Around the world, one hundred thousand synthetic chemicals are now on the market. Each year one thousand new substances are introduced, most of them without adequate testing and review. At best, existing testing facilities worldwide can test only five hundred substances a year. In reality, only a fraction of this number actually do get tested."

Somehow, we have managed to persuade ourselves that the human body generally does not mind how many different chemicals disrupt it simultaneously, as long as no single chemical is present in a dosage large enough to cause blatant harm all by itself. We seem to have overlooked the idea that maybe our bodies have only a limited amount of energy to devote to the task of dealing with the total quantity of toxic chemicals being encountered, not to mention dealing with X-rays and other kinds of ionizing radiation.

For illustrative purposes, let us take the round number 10,000. Let us suppose that most of us, through the air we breathe, the food we eat, the liquids we drink, the clothes we wear and the various substances that come into contact with our skin, are exposed to 10,000 different chemicals during a typical year. Let us further suppose that, for each of these 10,000 different chemicals, our exposure amounts to only one-ten-thousandth of the dose considered by the authorities to be dangerous. The key question then becomes: "Are we in danger of being chemically poisoned in the long run?" Your answer, dear Reader, is just as good as mine!

(I am aware that in certain cases a specific pair of harmful chemicals may be known to reinforce each other's harmfulness. But that does not contradict the general point I am making here.)

We also seem to think that synthetic chemicals, just like human beings, should be considered innocent until proven guilty. And here we are prone to play, or be fooled by, the "evidence" word game described in Chapter 11. As Sandra Steingraber has written, "Thus, many carcinogenic environmental contaminants likely remain unidentified, unmonitored, and unregulated. Too often, this lack of basic information is paraphrased as 'there is a lack of evidence of harm,' which in turn is translated as 'the chemical is harmless.'"

As a result of such intellectual and ethical lapses, we in the modern industrialized world have organized our economic activities in such a way that every year large numbers of adults, children, infants and developing fetuses are being sentenced to developmental disorders and/or to chronic illness and/or to premature death, all caused by inappropriate and excessive exposure to synthetic chemicals. That is surely the sad blunt truth, no matter how uncomfortable it makes us feel. And the irony underlying this huge health tragedy is that under conventional economics a significant portion of our total annual consumption of synthetic chemicals serves only to help confiscate decent productive opportunities from eager human beings (think, for example, of small-scale farming families and of people who are, or would like to become, skilled at building wooden boats) and to allocate corresponding work to nuclear and fossil-fuel energy. In short, in a great many cases synthetic chemicals tend to serve the needs of the economy rather than the needs of human beings.

Not all synthetic chemicals are harmful, at least not under realistic circumstances. But Hugh D. Crone, in offering examples of chemicals in various toxicity classes, makes the following observation with regard to the "hardly toxic (considered harmless)" toxicity class: "Not too many synthetic chemicals would fall [into] this class." In other words, caution is definitely in order.

Equally important, not all harmful substances fall into the category of synthetic chemicals. For example, the so-called "heavy metals", such as mercury (Hg) and lead (Pb), each consist not of a synthetic chemical compound but rather of one of the elements of chemistry's periodic table. Nevertheless, heavy metals too can cause severe health problems once they find their way into the human body in even quite small concentrations. So can various plant poisons, snakebite venoms and so on. So can various other naturally occurring chemical compounds, to one of which, as an example, I now turn.

Hydrogen sulphide

Known informally as "rotten egg gas", hydrogen sulphide (H_2S) is a nasty gas that is often encountered by the oil and gas industry, especially in association with deposits of natural gas. Everyone agrees that elevated concentrations of hydrogen sulphide in natural gas are very dangerous. In the words of Andrew Nikiforuk, written in 2001, "A [natural] gas deposit doesn't have to contain much H_2S to be lethal. Just 0.06 of 1 percent, or 600 parts per million (ppm), can kill a man." That being the case, ethical principles would seem to require a very high standard of care on the part of any industry dealing with hydrogen sulphide, even in situations where the concentrations are far below 600 ppm. Referring specifically to Alberta, however, Nikiforuk goes on to say, "But both industry and government argue that no conclusive body of scientific evidence supports the claim that small doses [i.e. doses far smaller than 600 ppm] of H_2S are harmful. Big Oil seems to be today where Big Tobacco was 15 years ago: deny, deflect, dismiss."

Put another way, hydrogen sulphide in small doses, i.e. in doses much smaller than 600 ppm, is considered to be innocent until proven guilty; and the burden of proof falls on those who question that innocence. Strictly speaking, I admit, burdens of proof are determined by legal systems and regulatory systems, not by economic systems. But under conventional economics, the financial pressure—upon legislatures as well as upon the oil industry—to throw caution to the winds becomes well-nigh irresistible. We see huge dollar signs in the oil patch and we tell ourselves, "Surely a little exposure to hydrogen sulphide on the

346

part of a few farmers or ranchers or oil-workers or aboriginal residents can't be allowed to come between this tremendous good fortune and ourselves. Let's not pussy-foot around. There is work to be done and money to be made!" Power indeed corrupts.

The hydrogen sulphide issue is still very much alive. A small news item in *The Globe and Mail* of March 8, 2014 stated that a major Canadian oil and gas company was "facing 11 environmental charges over the release of a potentially deadly gas near an aboriginal community in northern Alberta." The gas in question was hydrogen sulphide and the alleged release took place in August, 2012.

INF economics will change the habits of our oil and gas industry quite substantially. In particular, any and all hydrocarbons brought up to, or allowed to escape to, the surface of the Earth will be subject to the new fossil-fuel energy tax. The tax will be payable regardless of whether or not there exists at that moment (a) any market for those particular hydrocarbons or (b) any pipeline or storage system able to accommodate them. Flaring off (i.e. burning off) unwanted hydrocarbon gases will thus amount to flaring off unwanted money!

Conservation and prudence will start receiving top priority. Moreover, government regulations concerning pollution in the oil patch can be expected to be tightened up considerably as the general public becomes more and more aware of the trade-offs between (a) short-term oil wealth and (b) long-term health and safety and community morale.

Contrast all this with the oil patch under conventional economics. Andrew Nikiforuk, writing in 2001, noted: "To date the government of Alberta has introduced no legislation to force companies to conserve gas, to stop flaring, or to pay royalties on flared or vented gas." Gordon Laird, writing in 2002, added a detail of interest: "And, for all the affluence Alberta holds, some rural residents find themselves saving up for their own electronic equipment to monitor air quality because the province is short on resources."

The Third World

(The term "Third World" seems to have fallen out of favour. But I continue to use it because, like many other people, I feel that the word "develop", together with all its various other forms, has been inappropriately borrowed from biology and applied to economics. In my view, there is nothing "developed" about the industrialized countries of the world. They are simply industrialized and wealthy and have enormous energy appetites. By the same token, the so-called "underdeveloped" or "developing" countries, also known as "Third World" countries, are basically much poorer and less industrialized than Europe, North America, Japan and so on.)

The Third World is far too large a topic to be dealt with in one small section of one small chapter of one small book. All I wish to say here is that in my view a changeover from conventional economics to INF economics by the modern industrialized countries of the world would surround the Third World with a far more benign environment than is the case at present: a far more benign economic environment and also a far more benign physical environment. Accordingly, any modern industrialized country that decides to lead the way in making such a changeover would be behaving in a highly ethical manner towards the Third World.

Let us acknowledge the obvious. We in the First World set a terrible example for the poorer countries of the world. We waste so much, while they have to make do with so little. We treat the environment like a garbage dump, but we express great indignation when they follow suit. We use much of our technology for the purpose of disenfranchising our own human energy, and we thrust such technology onto the Third World almost as heedlessly as we thrust it upon ourselves.

Vandana Shiva has been particularly scathing in her criticism of the "maldevelopment" policies and technologies that the First World has been inflicting on the Third World, especially with regard to agriculture: "Since then, western patriarchy's highly energy-intensive, chemical-intensive, water-intensive and capital-intensive agricultural techniques for creating deserts out of fertile soils in less than one or two decades has spread rapidly across the Third World as agricultural development, accelerated by the green revolution and financed by international development and aid agencies." Is Shiva's finger here not pointing at least partly at conventional economics?

In addition to setting a bad example and offering inappropriate advice, we in the First World sometimes exploit the Third World shamelessly in order to obtain economic benefits for ourselves, notably in connection with arms sales and waste disposal. Under conventional economics, manipulative salesmanship is every bit as alive and well on the international scene as it is here at home.

Given the example we set, the advice we offer and the exploitation we perpetrate, those who grow up in the poorer countries of the world must at times find life to be deeply demoralizing. Put more positively, it is truly amazing that so many of the inhabitants of Africa, Asia and Latin America have remained honest, cheerful, hard-working and undaunted, in spite of everything. Imagine what the Third World might be like if we in the First World actually practised and cherished frugality and energy efficiency. Imagine the example that we would be setting. Imagine the appropriateness of the technological advice and assistance that we would be able to offer. Imagine the solidarity, the confidence, the optimism and the determination—on both sides.

But I must not get carried away. Utopias do not come easily. My basic point is simply that potential Third World benefits constitute an additional ethical reason why a First World country such as Canada might consider adopting INF economics.

Petrotyranny

In his book, *Petrotryanny*, John Bacher puts forward the argument that oil wealth and political dictatorship are often closely linked, as are oil wealth and war. We need to replace "the deadly trinity of oil, war and dictatorship," Bacher states, with "a holy trinity of peace, human rights and environmental sustainability." Few books underline more sharply the negative political aspects of the role played by oil wealth on the world stage.

As I see it, Bacher's argument brings us right back to the central fact that power corrupts and absolute power corrupts absolutely. Once a certain amount of corruption has already occurred, power in the form of oil wealth can easily get converted into power in the form of a repressive political dictatorship.

No one is seriously accusing Canada of being a dictatorship, despite certain flaws in our practice of democracy. But the essentially democratic nature of our system of government does not absolve us of ethical responsibility here. At any given time, Canadian oil companies may have made significant investments in the oil industry of one or more oil-rich but democracy-poor Third World countries. Such investments help to prop up the regime in question in two different ways: (a) by using Canadian capital and Canadian entrepreneurial talent to help convert the underground oil reserves of country X into cash and (b) by lending Canadian prestige and apparent Canadian approval to the political regime in power in country X.

In addition and in a more general way, Canadian acceptance of conventional economics helps the rest of the modern industrialized world maintain its belief, or at least its claim, that low prices for fossil-fuel energy are good for everyone. In actual fact, low prices bring about high levels of demand. And a high level of world demand is exactly what a dictatorial regime needs in order to convert large amounts of its oil reserves into cash and thus into revenue for itself. In short, the First World's ongoing acceptance of conventional economics makes it all too easy for a number of petrotyrannies to retain their grip on political power.

Canada is obviously not in a position to force any of the world's present petrotyrannies to democratize themselves. But ethical conduct is not about control. It is about doing what is right. And it is about setting a good example. With regard to the world's present petrotyranny blight in particular, INF economics will offer Canada a basic energy policy that is far more ethical, in my view, than our current choice.

Good intentions

Finally, let me return to the ethics of our present behaviour towards the Earth and the natural world. I offer the following thought from Thomas Berry: "If this assault on the earth were done by evil persons with destructive intentions, it would be understandable. The tragedy is that our economy is being run by persons with good intentions under the illusion that they are bringing only great benefits to the world and even fulfilling a sacred task on the part of the human community."

23

Getting There

"Social change will not come to us like an avalanche down the mountain. Social change will come through seeds growing in well prepared soil—and it is we, like the earthworms, who prepare the soil. We also seed thoughts and knowledge and concern. We realize there are no guarantees as to what will come up. Yet we do know that without the seeds and the prepared soil nothing will grow at all."

Ursula Franklin (1990, 1999)

"The lessons of life have taught us how little use a political democracy will be, however well-balanced it may appear in its internal structures and institutional functioning, if it is not constituted as the basis for an effective and real economic democracy and for a no less real and effective cultural democracy."

José Saramago (2010)

How can we get ourselves to deliberately turn away from the doctrine of perpetual economic growth and in its place adopt Intelligent National Frugality (INF) economics or something comparable?

Words

Let me begin with our verbal behaviour. In today's world, nothing is easier than to sprinkle our ordinary conversation with comments suggesting that we agree wholeheartedly with the doctrine of perpetual economic growth. Even if we do not actually believe in that doctrine, we often feel reluctant to make waves by voicing our reservations. "Yes, his new sports car goes like the wind!" we say with apparent admiration and possibly even apparent envy. Or, "That new copper mine? It's the best thing that has happened around here in ages." Or, "What you are seeing there, my friend, is the skyline of a world-class city."

Indeed, one tends to feel like a party-pooper if one withholds one's approval of the latest example of modern energy extravagance. Yet if we all profess to be in favour of such extravagance, how can we possible expect our society to grant national frugality a toehold? I am not suggesting that we should all go around spouting self-righteous condemnation of just about everything in the modern world. One need only remain silent or, if questioned, respond with a low-key comment such as: "I personally would not have done that," or "I see losses there as well as gains," or "That seems rather foolish to me."

Just as neither a pacifist nor a semi-pacifist would further their cause if they were to voice support for a war that they inwardly opposed, so we frugalists lose ground if we allow ourselves to utter platitudes suggesting that we too believe in perpetual economic growth. As Kenneth E. Boulding has noted, "In conversation, writing, and in the ordinary activity of daily life we are constantly communicating with others, and as a result of these communications their images of the world change ... [A]ll of us are teachers whether we like it or not, whether we get paid for it or not, or even whether we are conscious of it or not."

Deeds

What about our deeds, as distinct from our words? In our daily lives, can our behaviour as individuals have any influence on the direction in which society is heading? Here, there are no guarantees. Wisdom does not always triumph over folly. But the whole point about ethics and ethical behaviour is that one does the right thing because it is right, not because it will necessarily result in a happy outcome for all concerned. As Wendell Berry stated at a San Francisco conference in 1993, "As far as I can see the majority is not on our side. We're a small minority and we are losing. What we have to do may be impossible, and that makes no difference at all. Win or lose, our task is to work toward a less destructive economy."

We can work towards the goal of a less destructive economy (a) by trying to integrate our ideals into our daily lives, (b) by joining (or otherwise supporting) suitable organizations and movements and (c) by keeping in mind the basic fact that conventional economics is constantly trying to steer us in the wrong direction. I agree with Chris Hedges' statement: "We can, and should, live more simply, but it will not be enough if we do not radically transform the economic structure of the industrial world."

How can one live one's life in an ethical and frugal manner in a society whose economic philosophy and economic structure put pressure on everyone to play the "economic growth" game? The short answer, too short to be of much help, is that one does one's best.

For a longer answer, I offer the following thoughts.

In order to resist the siren call of conventional economics, it is necessary first and foremost to stop believing: to stop believing that our political leaders, our business and banking leaders and our mainstream economic "experts" have a fundamental understanding of what they are saying and doing and where they are leading us; to stop believing that perpetual economic growth is both possible and desirable; to stop believing that the combination of money and technology offers the surest solution to nearly all our social, economic and environmental problems; to stop believing that almost everything important has a dollar value; and to stop believing that an acceptable economic philosophy can blithely ignore both frugality and ethics.

Next, we will want to start believing in something else, something that need not be narrowly limited to INF economics. Other alternatives are being put forward all the time, alternatives that share with INF economics a focus on sensible and ethical approaches to national or local frugality. Regardless of which specific alternative each of us favours, primacy needs to be given to human effort, human intelligence, human creativity and excellence, human morale, human moderation, human generosity, human cooperation, and human concern for future generations and for the future of Gaia.

In my view, we do not have to divorce ourselves completely from fossil-fuel energy. But we have to stop putting the cart in front of the horse. We have to stop putting the voracious resource appetite of our national economy ahead of the real needs of the human beings whom the economy is supposed to serve.

As our esteemed descendant reminded us in Chapter 3, one of civilization's oldest ethical principles is that we should all try to leave the world a better place than we found it. Somehow, we moderns seem to have forgotten that idea. Or if we remember it, we pay it lip service only, pretending to ourselves that many of our most flagrantly wasteful and unsustainable activities will somehow provide benefits lasting far into the future.

We need to reintegrate that ethical principle into our personal philosophy of life. We will then make much better decisions regarding our own personal economic behaviour. We will see the ethical pointlessness of trying to "keep up with the Jones" and of playing endless games of one-upmanship with our neighbours and acquaintances. We will find it much easier to perceive and resist the manipulative aspects of modern advertising and modern marketing. In many situations, we will give serious consideration to using our own human energy, combined with hand tools, bicycles, horses and so on, rather than turning almost automatically to machines, gasoline, electricity and other energy slaves. We will pay attention to how far, how fast and how energy-efficiently we travel each day and each year. In short, we will try to use energy—and non-renewable resources in general—as wisely and sparingly as we reasonably can.

As long, however, as our society continues to saddle itself with conventional economics, we will probably all have difficulty in deciding how much to give up in our personal lives, no matter how convinced intellectually we might be of the attractiveness of intelligent frugality. Few people qualify as saints. Few people are willingly going to turn their backs on all mechanized travel, all modern entertainment, all modern conveniences and all modern electronics. Nor should they. But finding a personal balance between significant frugality and reasonable participation in the life of the modern world under conventional economics is never going to be easy. From time to time, we will all no doubt feel, and behave, somewhat like hypocrites.

Likelihood

This brings me to two questions about likelihood. How likely is it that we will succeed in getting our country to adopt INF economics or something comparable? And how much time is that achievement likely to take? The short answer in both cases is that no one knows. The future behaviour of large groups of human beings is essentially unpredictable, as is confirmed by numerous historical examples. All we can do is live our lives in the present with as much wisdom, hope and ethical courage as we can muster.

I like to think that the picture of the future painted for us by our esteemed descendant in Chapter 3 will turn out a few generations hence to be fairly accurate. But it might prove to be either wildly optimistic or laughably timid. Neither eventuality would in my view invalidate the arguments that I have been putting forward in support of our switching over quite soon to INF economics. If Chapter 3 turns out to be wildly optimistic, then our need for some form of national frugality today will in hindsight prove to have been all the more urgent. If on the other hand Chapter 3 turns out to be excessively timid, then frugality today will simply have kept available a great many options and resources for further adventures tomorrow.

A time of crisis

As I and many others see it, we are not living in ordinary times. We are living in a time of crisis, and not just a temporary economic or financial crisis. Various danger signs on the horizon indicate that if we in the modern industrialized countries of the world continue on our present course, we may be heading for disaster. In the worst case, much preventable suffering and many premature deaths could occur, with the most likely proximate cause being a shortage of food and potable water (where needed) and/or a shortage of adequate shelter (where needed). And we need to keep in mind that, in much of the world, adequate shelter includes both adequate indoor wintertime heat and an absence of prolonged unbearable indoor summertime heat.

I can imagine two different approaches by which we might nip this impending disaster in the bud.

The first approach is risky but might conceivably work: the "dribs and drabs" approach. Without any kind of overall plan or strategy, we would simply fight each individual economic fire as it broke out. In doing so, we would move jerkily and uncertainly, by dribs and drabs, towards a more frugal future. Warren Johnson examined this approach in his 1978 book *Muddling Toward Frugality*, and much of what I have written in the present book was said decades ago by Johnson.

One problem with the dribs and drabs approach is that it tends not to generate very much motivation, inspiration or enthusiasm. A hundred years from now, some future historian might write a fascinating book titled *How We Muddled Toward Peace in the Middle East*. But who right now wants to read something with a title like *Switching Over to Renewable Energy by Dribs and Drabs*? And what if one feels that by now it is too late for the dribs and drabs approach?

While pushing for a much more concerted approach, however, we should surely in the meantime accept and welcome every drib and every drab that comes our way, such as a carbon tax, no matter how small, or an attempt at promoting local food supplies, or a proposal for a more comprehensive interurban passenger rail service.

The second and more concerted approach would involve less uncertainty than the first but would require a great deal of determination and commitment on the part of a great many people. It would feature an explicit plan to adopt either Intelligent National Frugality or something comparable. Having thoroughly discussed and debated the matter beforehand, our whole society would perhaps formally place itself on an emergency footing for part or all of the transition period. Details of the planned transition would be democratically worked out as far in advance as reasonably possible.

With a certain amount of misgiving, knowing that warfare is always unspeakably violent and usually accomplishes nothing worthwhile, I see a parallel between (a) many of the changes that would come into play following a national declaration of frugality and (b) many of the changes that occurred in Canada and in its allies at the beginning of the Second World War: huge changes in almost everyone's attitude, changes in personal goals and plans, changes in economic priorities, changes in business behaviour and banking behaviour, changes in society's values. In such a situation, one expects a much greater focus on cooperation and concern for the greater good, and much less wondering, "What's in it for me?"

Conclusion

As I mentioned in the Introduction, all or nearly all the ideas in this book have been around for quite some time. Whether they end up germinating depends both upon their intrinsic validity and upon the wishes, beliefs, values, decisions and behaviour of large numbers of people. I hope that our collective wisdom and our collective ethical inspiration will soon be able to overwhelm our current love affair with the untrustworthy triad of foolish risk-taking, deliberate blindness and far too large a brigade of energy slaves. May we all help that happen!

###

Endnotes

General

1. For ease of reading, the text of this book deliberately omits subscript numbers that would refer to these endnotes. Nevertheless, the reader should have no difficulty in establishing the intended connections.

2. These endnotes are to be read in conjunction with the corresponding entries in the bibliography.

3. In some cases, the page number(s) for a particular quotation or reference may vary slightly from one edition of a book to another.

Modern Energy Extravagance

1. Flannery, p. 128.

2. Turner, p. 2.

3. Andersen, pp. 289 and 294.

4. Judt, p. 161.

5. Berry, Thomas (1999), p. 165.

6. Diamond, pp. 79-119 and 520-521.

7. Turner, p. 343.

8. Wright, pp. 57-64.

9. Tuchman, pp. 4 and 32.

10. Concerning originality, Georgescu-Roegen (Reprint 2013), p. xiii, comments as follows: "[P]ractically all works we usually

call our own represent only a few scoops of originality added on top of a mountain of knowledge received from others."

11. Lovelock (2006), p. 162.

12. Margulis (1998), pp. 1-2 and 113-128.

13. For comments by Lovelock concerning his collaboration with Margulis, see Lovelock (1995), pp. xviii, xix and 8.

Chapter 1: Human Energy

1. Azimov, p. 180 (in Bantam Pathfinder Edition).

2. Greenfield, p. 27.

3. Soddy (1983), p. 56.

4. Berry, Wendell (1977), p. 13.

5. Hawken et al. (1999), p. 55.

6. Orwell, pp. 220-221.

7. Turner, pp. 30-31.

Chapter 2: Intelligent National Frugality (INF) Economics

1. Hannon (1977), p. 51.

2. Daly (2011), p. 2.

3. Pielou, p. 213. Pielou here is explaining a hypothesis. She is not expressing any personal opinion on the matter.

Chapter 3: A Peek at the Future

1. A somewhat similar letter, albeit with significant differences, appears as Chapter 10 in Heinberg (2007).

2. Merchant, p. 295.

3. Orr, p. 151. Note also that Chapter 9 in McDaniel focuses on David Orr.

4. Johnson wrote on p. 130: "It will lead to the decentralization of the economy, to smaller scale technologies, to the repopulating of rural areas, and to reducing the overloading of our cities."

5. Essays by a number of critics of industrial agriculture appear in Kimbrell and in Wirzba.

6. Important suggestions concerning municipal zoning and municipal performance codes appear in Jacobs (2004), pp. 153-157.

7. Kunstler (2012), pp. 51-56, offers several reasons why in his opinion "the skyscraper is obsolete."

8. Illich, p. 12.

9. Berry, Thomas (1999), p. 7.

10. Klein, p. 466.

11. Of relevance to much of this whole chapter is the essay by Northrup and Lipscomb, even though they make no mention of anything resembling INF economics.

Chapter 4: Making the Tax Progressive

1. Brockway (1995), p. 152.

2. Roodman, p. 188.

3. Robertson, p. 81.

Chapter 5: Exports, Imports and Free Trade

1. Henderson, p. 179.

2. Daly (1996), p. 157.

3. Berry, Wendell (1995), p. 14.

4. Daly (1996), p. 150.

5. The essay "Flight" appears in Lopez, pp. 73-109. The quotation is on p. 85.

6. Hardin's 1968 essay "The Tragedy of the Commons" and his subsequent essay "Second Thoughts on 'The Tragedy of the Commons'" both appear in Daly and Townsend, pp. 127-143 and pp. 145-151.

Chapter 6: Lower Levels of Government

1. Roodman, pp. 120-121.

Chapter 7: The Transition

1. Foley, p. 302.

2. Hawken (1993), p. 180.

3. Smil (2008), p. 363.

4. McKibben (2010-2011), p. 40.

Chapter 8: Nuclear Power

1. Caldicott (1994), p. 32.

2. Franklin (1999), p. 125.

3. Smil (2003), p. 84.

4. Bertell, p. 53. Note also that Chapter 2 of Freeman focuses on Rosalie Bertell.

5. Cassedy and Grossman, p. 216.

6. Macy, p. 223.

Chapter 9: Environmental Overview

1. McMichael, p. 98.

2. Asimov and Pohl (1991), p. 89.

3. Dubos (1968), pp. xi-xii.

4. Jackson (1985), pp. 63-65.

5. McDaniel, p. 82.

6. Ehrlich and Ehrlich, p. 5.

7. Suzuki (1998), p. 110. For an interview of Suzuki by Farley Mowat, see Mowat, p. 169.

Chapter 10: Selected Energy-Environment Links

1. Brinkhurst and Chant, p. 41.

2. Nikiforuk (2010), p. 83.

3. May, p. 50. For an interview of May by Farley Mowat, see Mowat, p. 188. And for May's essay "Gaia Women", see Mowat, p. 247.

4. Suzuki (1994), pp. 32-33.

5. Roebuck, p. 75.

6. Smil (2003), p. 256.

7. Concerning the deliberate demolition of certain U.S. dams, see "The Penobscot's song" in *The Economist*, June 16th, 2012, p. 38.

8. Garrett, p. 426.

9. Logsdon, p. 83.

10. For a more in-depth look at the use of antibiotics on farm animals, see Levy, pp. 137-156. The quoted passages appear on p. 137 and p. 140.

11. Specter, p. 36.

12. André Picard, in "We will pay for antibiotic abuse" in *The Globe and Mail*, October 14th, 2014, p. A 17, states: "Globally, about 75% of antibiotics are consumed by farmed animals—chickens, cows, pigs, fish and so on. They are pumped into animals to make them bigger, fatter and more profitable. It's also cheaper to buy antibiotics than to invest in better hygiene for them—at least in the short term."

13. Roy, p. 60.

14. Berry, Wendell (2000), *Life is a Miracle*, pp. 43-44.

15. Berry, Wendell (2012), p. 215.

16. Lovelock (2006), p. 132.

Chapter 11: Global Warming and Global Climate Change

1. Smil (2001), p. 97.

2. Dotto, p. 299.

3. Henderson-Sellers, p. 104.

4. Regarding 395 and 280 ppmv, see Carey, p. 52. And for the two Carey quotations, see Carey, p. 54 and p. 52.

5. Lovelock (1991), p. 142.

6. Sagan, p. 145.

7. Caldicott (1994), p. 15.

8. Nikiforuk (2010), p. 151.

9. Lovelock (2006), pp. 90-104.

10. Concerning CCS in Weyburn, Sask. and off the coast of Norway, see Casey.

11. For a detailed description of the Lake Nyos problem, see Holloway.

12. Scott, p. 28.

13. Berry, Thomas (1999), p. 158.

Chapter 12: Why Not a Carbon Tax?

1. *The Economist*, June 16th, 2012, p. 13.

2. Kolbert (2014), p. 21.

Chapter 13: Economic Growth and Its Underlying Ideology

1. Mishan, p. 19.

2. Georgescu-Roegen (1976), p. xv. For an important essay titled "On Nicholas Georgescu-Roegen's Contributions to Economics: An Obituary Essay", see Daly (1996), Chapter 13, pp. 191-198.

3. Daly and Cobb (1994), pp. 63-64.

4. Some writers do use the word "perpetual" and/or the word "perpetually" in describing the unending nature of the economic growth pursued by conventional economics. See for example Ehrenfeld, p. 193; Heinberg (2007), p. 174; McNeill, p. 335; Nadeau, passim.

5. Dubos (1972), p. 228.

6. Abbey, p. 45.

7. Daly and Farley (2004), p. 16 and p. 20.

8. Jacobs (2000), pp. 54-55.

9. Sacks, p. 247.

10. Mirza is quoted in Perreaux.

11. Layton, p. 227.

12. Klein, p. 115.

13. Schumacher, p. 260 (in Harper Torchbook edition).

14. For depictions of the unpleasantness and the moral failings of several specific ideologies of the past, see Paris.

15. The Tennyson phrase appears in the poem "Charge of the Light Brigade".

16. Homer-Dixon (2000), p. 308.

17. Jacobs (2000), p. 147.

18. McNeill, p. 335 and p. 336.

19. Daly and Farley (2004), pp. xxvi, xxiv and xviii.

Chapter 14: Sustainability and Frugality

1. Kelly, pp. 23-24. This same essay by Kelly, titled "Creating an Ecological Economy", also appears in Tobias and Cowan.

2. Jackson (1996), p. 3.

3. George, p. 406: "[W]e may put the proposition into practical form by proposing To *abolish all taxation save that upon land values.*" Incidentally, one could argue—although I do not pursue the argument in my text—that INF economics essentially applies George's reasoning to the modern economic world. One could argue that modern land values depend upon two factors: a) the ongoing benefit accruing to the landowner via ownership of the surface rights and (b) the one-time-only benefit accruing to the landowner via ownership of the underground mineral rights and fossil-fuel resource rights. Under INF economics, the first factor would be taxed annually by municipal taxes on land (with buildings NOT being taxed), while the second factor would be taxed— just once—by the INF tax itself.

4. Dowie, p. 235.

5. Georgescu-Roegen (Reprint 2013), p. 304.

6. Veblen, p. 78.

7. McDaniel, p. 150.

Chapter 15: Supply, Demand and Price

1. Brockway (1993), p. 77.

2. Georgescu-Roegen, p. 33.

3. Illich, pp. 4, 6 and 10.

4. Nikiforuk (2012), p. 68.

5. Smil (2003), pp. 360-361.

Chapter 16: The Price of Fossil Fuels

1. Georgescu-Roegen (1976), p. 30 (footnote).

2. Daly (1991), p. 33.

3. Georgescu-Roegen (1976), p. 30.

Chapter 17: Efficiency

1. Ehrenfeld, p. 188.

2. Henderson, p. 63.

3. The term "negawatts" appears several times in Hawken, Lovins and Lovins (1999), p. 279.

4. For an unflattering examination of Alberta's upgraders (present and future), see Nikiforuk (2010), pp. 112-121.

5. For more on EROEI, see Heinberg (2007), p. 151, Heinberg (2015), pp. 12, 25-28 and 178, Homer-Dixon (2006), pp. 49-53, and especially Smil (2008), pp. 275-281 and 377.

6. Jacobs (2004), pp. 158-160.

Chapter 18: Productivity

1. Hawken, p. 88.

2. Franklin (2006), p. 248.

3. Brockway (1993), pp. 163-164.

4. Brockway (1993), p. 165.

5. For the Mander quote, see Mills, p. 154.

6. Cassedy and Grossman, p. 148.

7. Hawken, p. 69.

8. Carter, p. 45.

9. Half-truths can also be found in discussions about other kinds of productivity, notably agricultural productivity. Regardless of the kind of productivity involved, failure to acknowledge the other half of the truth is always to be deplored. For more on this point as regards industrial versus ecological agriculture, see Shiva (2004), pp. 131-138.

Chapter 19: Free Enterprise versus Socialism

1. Commoner (1971), p. 281.

2. Caldicott (1992), p. 14.

3. Illich, p. 4.

4. George, p. 167.

5. Henson, p. 239.

6. Berry, Wendell (2000), *Jayber Crow*, p. 247.

7. Ward (1966), p. 121.

8. Whitelegg, pp. 38-40.

9. Terry, p. 149.

10. Adam Smith, considered the intellectual father of free enterprise, wrote *The Wealth of Nations* in 1776.

11. De Graaf and Batker, p. 11.

12. Korten, p. 56.

Chapter 20: Advertising and Marketing

1. Galbraith (1958), p. 156.

2. Henderson, p. 153.

3. Boulding (1948), p. 594.

4. Galbraith (1973), p. 140.

5. More strongly than I myself, James Howard Kunstler has expressed doubts about the future of the Internet. His reasoning is based on his expectation of resource scarcities in the near future, rather than on the possible consequences of a deliberate embrace of intelligent national frugality. Those two scenarios, however, may not be mutually exclusive. In any case, here is one of Kunstler's key statements concerning the Internet: "The electronic server 'farms' composed of massed computers require too much electricity." See Kunstler (2012), p. 226.

6. Boulding (1948), p. 594: "There is a case for a certain amount of advertising, such as the purely informative advertising which is descriptive of the qualities and prices of commodities. This is a form of consumer education which is necessary if consumers are to make intelligent choices ..."

7. Carter, pp. 61-66 and 132-138. The words quoted appear on pp. 136 and 137.

Chapter 21: Money Creation

1. Soddy (1931), p. 108.

2. Trenn, p. 180.

3. Galbraith (1975), p. 208.

4. Galbraith (1975), pp. 54-55.

5. Daly (1996), p. 175. The whole of Daly's Chapter 12 (pp. 173-190) deals with Soddy's economic thought.

6. For further discussion of Soddy's economic thought, see Daly (1991), passim; Daly (1996), passim; Daly and Cobb, passim; Daly and Farley, passim; Foley, passim (including an epigraph on p. 7); Merricks, pp. 59-78); Nikforuk (2012), passim (including an epigraph on p. 158); Trenn, pp. 179-198.

7. Concerning incorrect descriptions in economics textbooks, McLeay et al. write, "This article explains how, rather than banks lending out deposits that are placed with them, the act of lending creates deposits—the reverse of the sequence typically described in textbooks" (p. 15) and "As with the relationship between deposits and loans, the relationship between reserves and loans typically operates in the reverse way to that described in some economics textbooks" (p. 15) and "This description [that we have just given] of the relationship between monetary policy and money differs from the description in many introductory textbooks ... " (p. 21).

8. Concerning constraints against excessive lending by the commercial banks, see McLeay et al., pp. 17-19.

9. Hixson, pp. 115-116.

10. Fisher, p. 20.

11. Daly and Farley (2004), p. 250.

12. All the Felix Martin quotations appear in Martin, pp. 245 and 253-255.

Chapter 22: Ethics and Economics

1. Ward (1976), p. 293.

2. Lovins, pp. 58 and 66.

3. For Lord Acton's quotation, see *The Oxford Dictionary of Quotations*, p. 1.

4. Illich, p. 6.

5. Foljambe, p. 68.

6. Referring to Lord Acton's famous dictum about corruption, James Lovelock has written, "He was thinking of political power, but it could be just as true of electricity." See Lovelock (1995), p. 164.

7. Cameron, pp. 18-19.

8. Chomsky, p. 149.

9. Colborn et al., p. 138.

10. Steingraber, p. 100.

11. Crone, p. 35, Table 4.2.

12. Nikiforuk (2001), pp. 20 and 22.

13. Nikiforuk (2001), p. 256.

14. Laird, p. 204.

15. Shiva (1989), p. 153.

16. Bacher, p. 17.

17. Berry, Thomas (1988), pp. 76-77.

Chapter 23: Getting There

1. Franklin (1999), p. 121.

2. Saramago, p. 27.

3. Boulding (1964), p. 194.

4. Wendell Berry is quoted in Mills, p. 6.

5. Hedges, p. 301.

6. Concerning wild optimism and laughable timidity, compare my comments with those of Bruce Hannon that appear at the end of his 1975 article on energy conservation: "If the [energy] crisis is real [and if we respond to that crisis by taking a 'conservative approach'], we shall be viewed by future generations as terribly wise; if the crisis is false [and if we still respond by taking a 'conservative approach'], we shall have had an instructive relief from the dancing mirage of our manifest destiny." See Hannon (1975), p. 102.

Bibliography

This bibliography lists only those titles and authors referred to in the main text or in the endnotes.

The Oxford Dictionary of Quotations. 1980 (Third edition 1979, reprinted with corrections). Oxford: Oxford University Press.

Abbey, Edward. 1971 (Paperback edition). *Desert Solitaire: A Season in the Wilderness*. New York: Ballantine Books. (First published 1968 by McGraw-Hill.)

Andersen, Hans Christian. No date. "The Emperor's New Clothes". In *Favorite Tales from Grimm and Andersen*. Toronto: Royce Publications. (This combined edition first published in English by Orbis Publishing, London, 1983.)

Asimov, Issac. 1962. *Life and Energy*. New York: Doubleday.

Asimov, Issac, and Frederick Pohl. 1991. *Our Angry Earth*. New York: Tom Doherty.

Bacher, John. 2000. *Petrotyranny*. Toronto: Dundurn Press.

Berry, Thomas. 1988. *The Dream of the Earth*. San Francisco: Sierra Club.

Berry, Thomas. 1999. *The Great Work: Our Way into the Future*. New York: Bell Tower.

Berry, Wendell. 1977. *The Unsettling of America: Culture and Agriculture*. San Francisco: Sierra Club.

Berry, Wendell. 1995. *Another Turn of the Crank*. Washington, D.C.: Counterpoint.

Berry, Wendell. 2000. *Life is a Miracle: An Essay against Modern Superstition*. Washington, D.C.: Counterpoint.

Berry, Wendell. 2000. *Jayber Crow*. Washington, D.C.: Counterpoint.

Berry, Wendell. 2012. *A Place in Time: Twenty Stories of the Port William Membership*. Berkeley, Calif.: Counterpoint.

Bertell, Rosalie. 1986. *No Immediate Danger? Prognosis for a Radioactive Earth*. Toronto: Women's Educational Press.

Boulding, Kenneth E. 1948 (Revised edition). *Economic Analysis*. New York: Harper. (Originally published 1941.)

Boulding, Kenneth E. 1964. *The Meaning of the Twentieth Century: The Great Transition*. New York: Harper and Row.

Brinkhurst, Ralph O., and Donald A. Chant. 1971. *This Good Earth: Our Fight for Survival*. Toronto: Macmillan.

Brockway, George P. 1993 (Revised edition). *The End of Economic Man: Principles of any Future Economics*. New York: W. W. Norton. (Original edition published 1991.)

Brockway, George P. 1995. *Economists Can Be Bad for Your Health: Second Thoughts on the Dismal Science*. New York: W. W. Norton.

Caldicott, Helen. 1992. *If You Love This Planet: A Plan to Heal the Earth*. New York: W. W. Norton.

Caldicott, Helen. 1994 (Revised edition). *Nuclear Madness: What You Can Do*. New York: W. W. Norton. (Original edition published 1978.)

Cameron, John. 1999. "More Precious Than Gold". *Resurgence*, no. 195, July/August 1999, pp. 18-19.

Carey, John. 2012. "Global Warming: Faster Than Expected?" *Scientific American*, November 2012, pp. 50-55.

Carson, Rachel. 1962. *Silent Spring*. Boston: Houghton Mifflin.

Carter, Charles. 1971 (Paperback edition). *Wealth: An Essay on the Purposes of Economics*. Harmondsworth, England: Penguin. (Originally published 1968.)

Casey, Allan. 2008. "Carbon Cemetery". *Canadian Geographic,* January/February 2008, pp. 56-66.

Cassedy, Edward S., and Peter Z. Grossman. 1998 (Second edition). *Introduction to Energy: Resources, Technology, and Society.* Cambridge: Cambridge University Press. (Original edition published 1990.)

Chomsky, Noam. 2000. "How Free is the Free Market?" In *Only Connect: Soil, Soul, Society: The Best of Resurgence Magazine,* selected by John Lane and Maya Kumar Mitchell. White River Junction, Vermont: Chelsea Green. (Originally published 1995.)

Colborn, Theo, Dianne Dumanoski, and John Peterson Myers. 1996. *Our Stolen Future: Are We Threatening Our Fertility, Intelligence, and Survival? A Scientific Detective Story.* New York: Dutton.

Commoner, Barry. 1971. *The Closing Circle: Nature, Man, and Technology.* New York: Alfred A. Knopf.

Commoner, Barry. 1992 (Paperback edition). *Making Peace with the Planet.* New York: The New Press.

Crone, Hugh D. 1986. *Chemicals and Society: A Guide to the New Chemical Age.* Cambridge: Cambridge University Press.

Daly, Herman E. 1991 (Second edition). *Steady-State Economics.* Washington, D.C.: Island Press. (First edition published 1977.)

Daly, Herman E. 1996. *Beyond Growth: The Economics of Sustainable Development.* Boston: Beacon Press.

Daly, Herman E. 2011. "What Should We Tax?" http://steadystate.org/what-should-we-tax/

Daly, Herman E., and John B. Cobb, Jr. 1994 (Second edition). *For the Common Good: Redirecting the Economy toward*

Community, the Environment, and a Sustainable Future. Boston: Beacon Press. (First edition published 1989.)

Daly, Herman E., and Joshua Farley. 2004. *Ecological Economics: Principles and Applications.* Washington, D.C.: Island Press.

Daly, Herman E., and Kenneth N. Townsend, editors. 1993. *Valuing the Earth: Economics, Ecology, Ethics.* Cambridge, Mass.: MIT Press.

De Graaf, John, and David K. Batker. 2011. *What's the Economy For, Anyway? Why It's Time to Stop Chasing Growth and Start Pursuing Happiness.* New York: Bloomsbury.

Diamond, Jared. 2005. *Collapse: How Societies Choose to Fail or Succeed.* New York: Viking Penguin.

Dotto, Lydia. 1999. *Storm Warning: Gambling with the Climate of Our Planet.* Toronto: Doubleday.

Dowie, Mark. 1995. *Losing Ground: American Environmentalism at the Close of the Twentieth Century.* Cambridge, Mass.: MIT Press.

Dubos, René. 1968. *So Human an Animal.* New York: Scribner's.

Dubos, René. 1972. *A God Within.* New York: Scribner's.

Eberhart, Mark E. 2007. *Feeding the Fire: The Lost History and Uncertain Future of Mankind's Energy Addiction.* New York: Harmony Books.

Ehrenfeld, David. 1993. *Beginning Again: People and Nature in the New Millennium.* New York: Oxford University Press.

Ehrich, Paul R., and Anne H. Ehrlich. 2013. "Can a Collapse of Global Civilization be Avoided?" Proceedings of the Royal Society B 280: 20122845. http://dx.doi.org/10.1098/rspb.2012.2845

Fisher, Irving. 1945 (Third edition). *100% Money*. New Haven, Conn.: City Printing Company. (Originally published 1935 in New York by Adelphi.)

Flannery, Tim. 2010. *Here on Earth: A Natural History of the Planet*. New York: Atlantic Monthly Press.

Foley, Gerald. 1981 (Second edition). *The Energy Question*. Harmondsworth, England: Penguin. (Originally published 1976.)

Foljambe, Alan. 2012. "Nothing is Absolute". *Popular Woodworking Magazine*, June 2012, p. 68.

Franklin, Ursula M. 1999 (Revised edition). *The Real World of Technology*. Toronto: House of Anansi. (First edition published 1990.)

Franklin, Ursula M. 2006. *The Ursula Franklin Reader: Pacifism as a Map*. Toronto: Between The Lines.

Freeman, Leslie J. 1981. *Nuclear Witnesses: Insiders Speak Out*. New York: W. W. Norton.

Galbraith, John Kenneth. 1958. *The Affluent Society*. Boston: Houghton Mifflin.

Galbraith, John Kenneth. 1973. *Economics and the Public Purpose*. Boston: Houghton Mifflin.

Galbraith, John Kenneth, 1975. *Money: Whence It Came, Where It Went*. Boston: Houghton Mifflin.

Garrett, Laurie. 1995 (Paperback edition). *The Coming Plague: Newly Emerging Diseases in a World Out of Balance*. New York: Penguin. (First published 1994 by Farrar, Straus and Giroux.)

George, Henry. 1929 edition. *Progress and Poverty: An Inquiry into the Cause of Industrial Depressions and of Increase of Want with Increase of Wealth: The Remedy*. New York: Modern Library. (Originally published 1879.)

Georgescu-Roegen, Nicholas. 1976. *Energy and Economic Myths: Institutional and Analytical Economic Essays.* Elmsford, New York: Pergamon.

Georgescu-Roegen, Nicholas. Reprint 2013. *The Entropy Law and the Economic Process.* Cambridge, Mass.: Harvard University Press. (Originally published 1971.)

Greenfield, Susan. 1997. *The Human Brain: A Guided Tour.* New York: Basic Books.

Hannon, Bruce. 1975. "Energy Conservation and the Consumer". *Science,* vol. 189, no. 4197, July 11th, 1975, pp. 95-102.

Hannon, Bruce. 1977. "Energy, Labor, and the Conserver Society". *Technology Review,* vol. 79, no. 5, (March/April 1977), pp. 47-53.

Hawken, Paul. 1993. *The Ecology of Commerce: A Declaration of Sustainability.* New York: HarperBusiness.

Hawken, Paul, Amory Lovins, and L. Hunter Lovins. 1999. *Natural Capitalism: Creating the Next Industrial Revolution.* Boston: Little, Brown.

Hedges, Chris. 2013. *The World As It Is: Dispatches on the Myth of Human Progress.* New York: Nation Books.

Heinberg, Richard. 2007. *Peak Everything: Waking Up to the Century of Declines.* Gabriola Island, B.C.: New Society.

Heinberg, Richard. 2015. *Afterburn: Society Beyond Fossil Fuels.* Gabriola Island, B.C.: New Society

Henderson, Hazel. 1996. *Building a Win-Win World: Life Beyond Global Economic Warfare.* San Francisco: Berrett-Koehler.

Henderson-Sellers, Ann. 1994. "Numerical Modelling of Global Climates". In *The Changing Global Environment,* edited by Neil Roberts. Cambridge, Mass.: Blackwell.

Henson, Dave. 2002. "The End of Agribusiness: Dismantling the Mechanisms of Corporate Rule". In *The Fatal Harvest Reader*, edited by Andrew Kimbrell (see below), pp. 225-239.

Hixson, William F. 1997. *It's Your Money*. Toronto: COMER Publications.

Holloway, Marguerite. 2000. "The Killing Lakes". *Scientific American,* July 2000, pp. 92-99.

Homer-Dixon, Thomas. 2000. *The Ingenuity Gap: Can We Solve the Problems of the Future?* Toronto: Alfred A. Knopf Canada.

Homer-Dixon, Thomas. 2006. *The Upside of Down: Catastrophe, Creativity, and the Renewal of Civilization*. Toronto: Alfred A. Knopf Canada.

Illich, Ivan. 1974. *Energy and Equity*. New York: Harper and Row.

Jackson, Wes. 1985 (New edition). *New Roots for Agriculture*. Lincoln, Nebraska: University of Nebraska Press. (Originally published 1980.)

Jackson, Wes. 1996 (Paperback edition). *Becoming Native to This Place*. Washington, D.C.: Counterpoint. (Originally published 1994 by the University of Kentucky Press.)

Jacobs, Jane. 2000. *The Nature of Economies*. Toronto: Random House Canada.

Jacobs, Jane. 2004. *Dark Age Ahead*. Toronto: Random House Canada.

Johnson, Warren A. 1978. *Muddling Toward Frugality: A Blueprint for Survival in the 1980s*. Boulder, Colorado: Shambhala. (Reprinted 2010 with a different subtitle. Westport, Conn.: Easton Studio Press.)

Judt, Tony. 2010. *Ill Fares the Land*. New York: Penguin.

Kelly, Petra K. 1994. *Thinking Green: Essays on Environmentalism, Feminism and Nonviolence*. Berkeley, Calif.: Parallax Press.

Kimbrell, Andrew, editor. 2002. *The Fatal Harvest Reader: The Tragedy of Industrial Agriculture*. Washington, D.C.: Island Press.

Klein, Naomi. 2014. *This Changes Everything: Capitalism vs. The Climate*. Toronto: Alfred A. Knopf Canada.

Kolbert, Elizabeth. 2006. *Field Notes from a Catastrophe: Man, Nature, and Climate Change*. New York: Bloomsbury.

Kolbert, Elizabeth. 2014. "Comment Rough Forecasts". *The New Yorker*, April 14th, 2014, pp. 21-22.

Korten, David C. 2001. "The Civilising of Global Society". In *Fiesta Review: Number 1*, edited by Richard Douthwaite and John Jopling. Dublin: FEASTA.

Kunstler, James Howard. 2005. *The Long Emergency: Surviving the End of Oil, Climate Change, and Other Converging Catastrophes of the Twenty-First Century*. New York: Atlantic Monthly Press.

Kunstler, James Howard. 2012. *Too Much Magic: Wishful Thinking, Technology, and the Fate of the Nation*. New York: Atlantic Monthly Press.

Laird, Gordon. 2002. *Power: Journeys Across an Energy Nation*. Toronto: Penguin Viking.

Layton, Jack. 2004. *Speaking Out: Ideas That Work for Canadians*. Toronto: Key Porter.

Levy, Stuart B. 1992. *The Antibiotic Paradox: How Miracle Drugs are Destroying the Miracle*. New York: Plenum Press.

Logsdon, Gene. 1994. *At Nature's Pace: Farming and the American Dream*. New York: Pantheon.

Lopez, Barry. 1998. *About This Life: Journeys on the Threshold of Memory.* New York: Alfred A. Knopf.

Lovelock, James. 1991. *Healing Gaia: Practical Medicine for the Planet.* New York: Harmony Books.

Lovelock, James. 1995 (Revised edition). *The Ages of Gaia: A Biography of Our Living Earth.* New York: W. W. Norton. (Originally published 1988.)

Lovelock, James. 2006. *The Revenge of Gaia: Earth's Climate Crisis and the Fate of Humanity.* New York: Basic Books.

Lovins, Amory B. 1977. *Soft Energy Paths: Toward a Durable Peace.* San Francisco: Friends of the Earth. Cambridge, Mass.: Ballinger.

Macy, Joanna. 1991. *World As Lover, World As Self.* Berkeley, Calif.: Parallax Press.

Margulis, Lynn. 1998. *Symbiotic Planet: A New Look at Evolution.* New York: Basic Books.

Martin, Felix. 2013. *Money: The Unauthorized Biography.* New York: Alfred A. Knopf.

May, Elizabeth. 1998. *At the Cutting Edge: The Crisis in Canada's Forests.* Toronto: Key Porter.

McDaniel, Carl N. 2005. *Wisdom for a Livable Planet.* San Antonio, Texas: Trinity University Press.

McKibben, Bill. 2010. *Eaarth: Making a Life on a Tough New Planet.* Toronto: Alfred A. Knopf Canada.

McKibben, Bill. 2010-11. "A Progressive Interview with Bill McKibben". *The Progressive,* vol. 74, no. 12/01, December 2010/January 2011, p. 39.

McLeay, Michael, Amar Radia, and Ryland Thomas. 2014. "Money Creation in the Modern Economy". *Bank of England Quarterly Bulletin 2014 Q1,* pp. 14-27.

McMichael, A. J. 1993. *Planetary Overload: Global Environmental Change and the Health of the Human Species.* Cambridge: Cambridge University Press.

McNeill, J. R. 2000. *Something New Under the Sun: An Environmental History of the Twentieth-Century World.* New York: W. W. Norton.

Merchant, Carolyn. 1980. *The Death of Nature: Women, Ecology and the Scientific Revolution.* San Francisco: Harper & Row.

Merricks, Linda. 1996. "Frederick Soddy: Scientist, Economist and Environmentalist—An Examination of his Politics". *Capitalism Nature Socialism,* vol. 7, no. 4, December 28th, 1996, pp. 59-78.

Mills, Stephanie, editor. 1997. *Turning Away From Technology: A New Vision for the 21st Century.* San Francisco: Sierra Club.

Mishan, E. J. 1969. *The Costs of Economic Growth.* Harmondsworth, England: Penguin. (First published 1967 by Staples Press.)

Mowat, Farley. 1990. *Rescue the Earth! Conversations with the Green Crusaders.* Toronto: McLelland & Stewart.

Nadeau, Robert L. 2003. *The Wealth of Nature: How Mainstream Economics Has Failed the Environment.* New York: Columbia University Press.

Nikiforuk, Andrew. 2001. *Saboteurs: Wiebo Ludwig's War Against Big Oil.* Toronto: Macfarlane Walter & Ross.

Nikiforuk, Andrew. 2010 (Revised edition). *Tar Sands: Dirty Oil and the Future of a Continent.* Vancouver, B.C.: Greystone Books, and David Suzuki Foundation.

Nikiforuk, Andrew. 2012. *The Energy of Slaves: Oil and the New Servitude.* Vancouver, B.C.: Greystone Books, and David Suzuki Foundation.

Northrup, Benjamin E., and Benjamin J. Bruxvoort Lipscomb. 2004. "Country and City: The Common Vision of Agrarians and New Urbanists". In *The Essential Agrarian Reader*, edited by Norman Wirzba (see below), pp. 191-211.

Orr, David W. 1994. *Earth in Mind: On Education, Environment, and the Human Prospect.* Washington, D.C.: Island Press.

Orwell, George. 2003 (Paperback edition). *Down and Out in Paris and London.* London: Penguin. (First published 1933 by Victor Gollancz.)

Paris, Erna. 2000. *Long Shadows: Truth, Lies and History.* Toronto: Alfred A. Knopf Canada.

Perreaux, Les. 2011. "Shutdowns of Rusty Bridges Cause Chaos in Montreal". *The Globe and Mail,* June 18th, 2011, p. A 13.

Pielou, E. C. 2001. *The Energy of Nature.* Chicago: University of Chicago Press.

Robertson, James. 2001. "Sharing the Value of Common Resources through Taxation and Public Expenditure". *Fiesta Review: Number 1*, edited by Richard Douthwaite and John Jopling. Dublin: FEASTA.

Roebuck, B. D. 1999. "Elevated Mercury in Fish as a Result of the James Bay Hydroelectric Development: Perception and Reality". In *Social and Environmental Impacts of the James Bay Hydroelectric Project*, edited by James F. Hornig. Montreal and Kingston: McGill-Queen's University Press.

Roodman, David Malin. 1998. *The Natural Wealth of Nations: Harnessing the Market for the Environment.* New York: W. W. Norton.

Roy, Arundhati. 1999. *The Cost of Living.* Toronto: Vintage Canada.

Sacks, Oliver. 2001. *Uncle Tungsten: Memories of a Chemical Boyhood.* Toronto: Alfred A. Knopf Canada.

Sagan, Carl. 1997. *Billions and Billions: Thoughts on Life and Death at the Brink of the Millennium.* New York: Random House.

Saramago, José. 2010. *The Notebook.* London: Verso.

Schumacher, E. F. 1973. *Small is Beautiful: Economics as if People Mattered.* New York: Harper & Row.

Scott, Dayna N. 2001. "Looking for Loopholes". *Alternatives Journal,* vol. 27, no. 4, Fall 2001, pp. 22-29.

Shiva, Vandana. 1989. *Staying Alive: Women, Ecology and Development.* London: Zed Books.

Shiva, Vandana. 2004. "Globalization and the War against Farmers and the Land". In *The Essential Agrarian Reader,* edited by Norman Wirzba (see below), pp. 121-139.

Smil, Vaclav. 2001 (Paperback edition). *Cycles of Life: Civilization and the Biosphere.* New York: Scientific American Library. (Original edition published 1997.)

Smil, Vaclav. 2003. *Energy at the Crossroads: Global Perspectives and Uncertainties.* Cambridge, Mass.: MIT Press.

Smil, Vaclav. 2008. *Energy in Nature and Society: General Energetics of Complex Systems.* Cambridge, Mass.: MIT Press.

Smith, Adam. 1937 edition. *The Wealth of Nations.* New York: Modern Library. (Originally published 1776.)

Soddy, Frederick. 1983 (New printing). *Wealth, Virtual Wealth, and Debt: The Solution of the Economic Paradox.* Hawthorne, Calif.: Omni Publications. (Originally published 1926 in London by George Allen & Unwin.)

Soddy, Frederick. 1931. *Money versus Man: A Statement of the World Problem from the Standpoint of the New Economics.* London: Elkin Mathews & Marrot.

Specter, Michael. 2012. "Germs are Us". *The New Yorker,* October 22nd, 2012.

Steingraber, Sandra. 1997. *Living Downstream: An Ecologist Looks at Cancer and the Environment.* Reading, Mass.: Addison-Wesley.

Suzuki, David. 1994. *Time to Change: Essays.* Toronto: Stoddart.

Suzuki, David. 1998. *Earth Time: Essays.* Toronto: Stoddart.

Terry, Roger. 1995. *Economic Insanity: How Growth-Driven Capitalism is Devouring the American Dream.* San Francisco: Berrett-Koehler.

Tobias, Michael, and Georgianne Cowan, editors. 1994. *The Soul of Nature: Celebrating the Spirit of the Earth.* New York: Continuum.

Trenn, Thaddeus J. 1986. "The Central Role of Energy in Soddy's Holistic and Critical Approach to Nuclear Science, Economics, and Social Responsibility". In *Frederick Soddy (1877-1956): Early Pioneer in Radiochemistry,* edited by George B. Kauffman. Dortrecht, The Netherlands: D. Reidel.

Tuchman, Barbara W. 1997 edition. *The March of Folly: From Troy to Vietnam.* London: Abacus. (First published 1984.)

Turner, Chris. 2011. *The Leap: How to Survive and Thrive in the Sustainable Economy.* Toronto: Random House Canada.

Veblen, Thorstein. 1994 edition. *The Theory of the Leisure Class.* New York: Penguin. (Originally published 1899.)

Ward, Barbara. 1966. *Spaceship Earth.* New York: Columbia University Press.

Ward, Barbara. 1976. *The Home of Man.* Toronto: McLelland & Stewart.

Whitelegg, John. 2002. "Flights of Fancy". *Resurgence,* no. 211, March/April 2002, p. 38.

Wirzba, Norman, editor. 2004 (Paperback edition). *The Essential Agrarian Reader: The Future of Culture, Community, and the Land.* Washington, D.C.: Shoemaker & Hoard. (Originally published 2003 by University Press of Kentucky).

Wright, Ronald. 2004. *A Short History of Progress.* Toronto: House of Anansi.

About The Author

Sometimes I think that a book like this should be written anonymously. That would force all readers and critics to focus on the book's contents rather than on the author's biographical details. On the other hand, we all like to know at least a little about the people with whom we interact. So I will state here that I am male and that I was born (in 1942), raised and educated in Ontario, Canada. For several years in the 1960s and 1970s, I lived and worked outside of North America. Since then, I have been living in Canada, mostly in rural eastern Ontario. I have worked at various jobs and on various projects over the years. In addition, since at least the 1970s, I have been thinking about and reading about the ideas discussed in this book. I have also spent countless hours putting my own thoughts down on paper, but no publishing company has ever agreed to publish them.

Even though the frugality dealt with in this book relates primarily to the business world and to government, the reader might be curious as to whether I have incorporated much frugality into my personal life. The answer is both "yes" and "no".

On the "yes" side, my wife and I have chosen to have no television set or dishwasher or clothes dryer in our house. Our refrigerator is similar to the one described in Chapter 3, our small, rural, super-insulated house is heated with wood, we have a composting toilet, and our highly efficient washing machine combines top-loading with a horizontal axis. We generate a portion of our electricity from our own solar panels and we buy the rest—which comes to us via the normal grid—from a commercial company that uses renewable sources for all the electricity it supplies to the grid. Much of the time, we use hand tools, wheelbarrows, clotheslines and human muscles.

On the "no" side, however, we drive our seven-year-old car long distances every year. Being a hybrid, it gets excellent fuel economy, provided that it is driven gently and conservatively. But clearly the combination of even excellent fuel economy with long driving distances does not constitute frugality. There was a five-year period in the early 1970s when, living in large cities, I relied mostly on public transportation and owned no motor vehicle at all. But under what I call conventional economics, I have never managed to get a firm grip on all aspects of personal frugality simultaneously. That in itself has taught me a lot.

In any case, let me emphasize that this book is not about me or about any other private individual. It is about Intelligent National Frugality, about how it might be achieved and about why I consider it to be a worthy goal.

John M. Braden